Texts in Philosophy

Volume 29

Carrollian Notes

Volume 19
History and Philosophy of Physics in South Cone
Roberto A. Martins, Guillermo Boido, and Víctor Rodríguez, eds.

Volume 20
History and Philosophy of Life Sciences in South Cone
Pablo Lorenzano, Lilian Al-Chueyr Pereira Martins, and Anna Carolina K. P. Regner, eds.

Volume 21
The Road Not Taken. On Husserl's Philosophy of Logic and Mathematics
Claire Ortiz Hill and Jairo José da Silva

Volume 22
The Good, the Right & the Fair – an introduction to ethics
Mickey Gjerris, Morten Ebbe Juul Nielsen, and Peter Sandøe

Volume 23
The Normative Structure of Responsibility. Law, Language, Ethics
Federico Faroldi

Volume 24
Karl Popper. A Centenary Assessment. Volume I. Life and Times, and Values in a World of Facts
Ian Jarvie, Karl Milford and David Miller, eds

Volume 25
Karl Popper. A Centenary Assessment. Volume II. Metaphysics and Epistemology
Ian Jarvie, Karl Milford and David Miller, eds

Volume 26
Karl Popper. A Centenary Assessment. Volume III Science
Ian Jarvie, Karl Milford and David Miller, eds

Volume 27
Unorthodox Analytic Philosophy
Guillermo E. Rosado Haddock

Volume 28
Quantum Heresies.
Kent A. Peacock, with a foreword by James Robert Brown

Volume 29
Carrollian Notes
George Englebretsen

Carrollian Notes

George Englebretsen

With a preface by

Francine F. Abeles

ISBN 978-1-84890-375-3

College Publications
Scientific Director: Dov Gabbay
Managing Director: Jane Spurr

http://www.collegepublications.co.uk

Original cover design by Laraine Welch

Finally, …

here's one for **Maëve**

CONTENTS

Preface

Now translated into 174 languages, Lewis Carroll's Alice in Wonderland published in 1865 has been delighting readers for generations. More than thirty years later, Carroll produced a sequel, Through the Looking Glass, with a cast of characters and situations revolving around them as inventive as in the first 'Alice' book.

Are Tweedledum and Tweedledee making sense or nonsense in their conversation,

> "I don't know what you're thinking about" said Tweedledum; "but it isn't so nohow" "Contrariwise" continued Tweedledee, "if it was so, it might be; and if it were so, it would be; but as it isn't, it ain't. That's logic."

The popularity of the two 'Alice' books cast a long shadow on its author so that when his two books on logic, The Game of Logic and Symbolic Logic, Part I appeared, reviewers dismissed their serious nature. In the academic world the 'Alice Effect' colored the lenses of professors too.

In the 20th century Dover republished these two books, also Carroll's book on geometry, Euclid and His Modern Rivals. These editions reinforced the 'Alice Effect' so that the serious issues Carroll presented in the two articles that he wrote for Mind, an important journal in logic and philosophy, "The Barbershop Paradox" and "What the Tortoise Said to Achilles" were not adequately addressed by scholars. Had Dover chosen to republish Dodgson's 1888 Curiosa Mathematica, Part I: A New Theory of Parallels, an entirely different view of Charles Dodgson, aka Lewis Carroll, Mathematical Lecturer at Christ Church in the University of Oxford might well have resulted.

Beginning in the 1970s a small group of scholars took up the task of examining the entire range of Dodgson's work in algebra, cryptography, geometry, logic, trigonometry, and voting theory. In the vanguard of this group was the logician and philosopher, George Englebretsen, whose essays, notes, and reviews from 1974 onward are the subject of this important volume. And a new piece, Lewis Carroll's Almost Diagrammatic Logic Notation", written with the noted Carroll scholar, Amirouche Moktefi, makes its debut here.

Dodgson was an unusual logician, both a traditionalist in his views of what logic was about, and a modernist in the mechanisms of expression that he worked with. A symbolist, he was a follower of George Boole who showed that the syllogism, the primary method of reasoning in logic from Aristotle's time through the early part of the twentieth century, could be treated algebraically.

Carroll was also a populist, believing that elementary logic could and should be taught to young people, especially if it were presented as a game as he did in his first book, The Game of Logic. This book also included a serious logical gem, his diagrammatic method to solve syllogisms. Compared to Venn diagrams, Carroll's set diagrams are just as easy to construct, they are self-similar, and they can represent an arbitrary number of sets.

If the reader of this volume is willing to invest the time, a cornucopia of logical jewels awaits.

Francine F. Abeles, Professor Emerita

Departments of Mathematics and of Computer Science

Kean University

Union, NJ, USA

Introduction

This is a collection of essays, notes, reviews, etc. published from 1974 to the present. The majority of these appeared in journals published by the Lewis Carroll Society (UK). I have appended brief notes (Notes 2021) to most of the early ones in order to indicate something of the contents of each and my take on them now, many years later. As a philosopher and logician, I've written often about Carroll's ideas and contributions to logic. And, of course, like so many philosophers and logicians (and many others as well), I was introduced from an early age to the wildly enchanting, genius of his fiction, and, like them, I've often quoted Carroll (especially from the *Alice* books) whenever I had even the slightest opportunity to do so.

The serious, systematic study of logic (in the West) began in the 4th century bce when Aristotle's explorations and inquiries into the fundamental principles of reasoning resulted in the logic of syllogisms (syllogistic logic). Over the next several centuries, that logic (with a variety of emendations) survived, first in the classical Greek world, then in the Arabic world's recovery and preservation of it from the 9th to 12th centuries, its recovery during the high middle ages by European Scholastics, and finally preserved in the logic of the 17th to 19th century. That logic, the logic that dominated scholarship from Aristotle to the end of the 19th century, has been generally known as "syllogistic" or "traditional" logic. The logic that Lewis Carroll knew and sought to critique and to improve in his own work was traditional logic. Carroll was, as he was in most things, a traditionalist.

By training and professionally, Charles Dodgson (far better known by his *nom de plume* as Lewis Carroll) was a mathematician. In fact, many of his contemporaries, George Boole, Agustus De Morgan, John Venn, John Cook Wilson, and Hugh MacColl, were mathematicians who, like Carroll, made major contributions to traditional logic. As logicians, they tended to look at logic as a kind of mathematics, specifically, a kind of algebra. They came to be known as the "British algebraic logicians." Ironically, at the very time that Carroll was engaged in his most important logical investigations and innovations in the 1880s and '90s, traditional logic's pride of place was, in very short order, being taken over by a new group of logicians. Most prominent among these was the German mathematician and logician, Gottlob Frege (along with the American Charles Sanders Peirce). The new logic was non-traditional, aimed not at providing a systematic, formal template of how people actually reason (as with Aristotle), but rather a new, contrived formal logical language, that could serve as the foundation of mathematics, not simply a branch of mathematics (as with the algebraic logicians). The dominant logic today is this "mathematical" logic.

A number of characteristics of the ways logical investigations were carried out by Carroll and his cohort were lost or diminished after the Fregean revolution. For example, with the exception of Peirce, mathematical logicians eschewed the traditional interest in developing diagrammatic methods for use in logical tasks. They took the logic of unanalyzed statements to be primary, relative to the logic of analyzed statements. They adopted a narrow notion of logical negation, treated logical quantity in a restricted way, and, most importantly and fundamentally, replaced the traditional account of logical form in terms of ordinary expressions combined by logical copulae with a radically different account in the familiar mathematical terms of *functions* and *arguments*.

The revolution that replaced traditional with mathematical logic was complete by the time Bertrand Russell and Alfred North Whitehead published their three-volume *Principia Mathematica* (1910-1913). Only rarely does one find the teaching of traditional logic in colleges or universities today (Peter Geach even called such places "colleges of unreason"). Of course, traditional logic, especially syllogistic logic, still enjoys some historical interest. But the general consensus is that whatever is of value there is nothing but a fragment of the new logic and is best treated in the terms that are now canonical. So why would anyone today be interested in what Lewis Carroll (who once referred to himself as an "obscure Writer on Logic, towards the end of the Nineteenth Century") wrote about logic?

Well, for one thing, Carroll was a wonderful writer, adept at the use of puns, puzzles, and tricks; and he often made use of that style when writing on logic matters. This is especially evident in his most important logical works, *Symbolic Logic*, *The Game of Logic*, and his two papers published in *Mind* ("What the Tortoise Said to Achilles" and "A Logical Paradox"). Carroll was particularly inspired by a belief that the elements of logic could be taught to children as long as it could be done in an appropriately accessible language and, especially, if it could be made fun. Thus he made the process, quite literally, a *Game*. And, naturally, he made use of many colorful, humorous, often silly examples to be used for exercises. Yet, for all that, logic was not merely a game for Carroll. His serious interest in logic was long-standing (as can be evidenced in his diaries. Among his important contributions to logic are his use of notational innovations, such underscoring of eliminable terms and subscribed numerals indicating negation, as aids for the decision process (determining the formal validity/invalidity of syllogisms), the introduction of his "method of trees" showing graphically how a valid argument can be shown to be valid by establishing that the set of premises along with the denial of the conclusion leads to a contradiction (it would be another three decades before modern logicians reinvented this device), his method of diagrams, which effectively extends the range of Venn-like diagrams, and, of course, his two papers in *Mind* that challenged not only his logical contemporaries but logicians ever since to think more carefully and deeply about the very idea of deduction.

Though I've had a few negative critical reactions to some of Carroll's logical claims, for the most part, my study of his work on logic over the past half-century has led me to a positive appreciation for many of the things he got right and for his various logical innovations. Much of what I've learned and explored in Carroll's work on logic is found in the notes, essays, and reviews to be found in this collection. The items collected here are arranged in chronological order. Needless to say, many of the pieces from the '70s and '80s would certainly be done more carefully or more knowledgeably were I to write them today. Also, I would not only correct some spelling and grammatical missteps, I would now be much more cautious in my judgments about Carroll as a logician. I would also be more careful in my deployment of personal pronouns. As I hope, and the reader should expect, later entries tend to reflect an increasingly substantial, deeper understanding of Carroll's many contributions to logic, contributions that are still valuable and worthy of the attention of today's community of logicians, philosophers of logic, and historians of logic.

Finally, this seem an appropriate place to express my deep and sincere appreciation to two Carroll scholars and friends, Fran Abeles and Amirouche Moktefi, who have aided and supported my own work in various ways over so many years. I thank Fran for her generosity in providing the Preface for the present work, and I thank Amirouche for his patience and expertise in co-authoring two of the items collected here. I also want to thank all the editors of the publications of the Lewis Carroll Society for the unfailing indulgence with which they treated me and my "obscure" works on logic. I also want to acknowledge and thank the following for their agreement to my use of much of the material collected here: De Gruyter Publishing; George Vanderburgh, Publisher, The Battered Silicon Dispatch Box; the Editorial Board of the Lewis

Carroll Society; the editors of the *South American Journal of Logic*; Springer Birkhaüser Publishing.

The Collection

1 The Tortoise, the Turtle, and Deductive Logic

Jabberwocky, (3), 1974, pp. 11-1

Only the most naive have ever doubted that the jokes which Charles Dodgson wrote as Lewis Carroll had very serious, sincere, and often philosophical points (cf. L. Wittgenstein, *The Philosophical Investigations*, New York, 1953, Sec. 111). Many of his jokes were grammatical or logical. In the *Alice* books Carroll makes literally hundreds of jokes, all designed, consciously or not, to provide lessons in mathematics, philosophy, linguistics, and especially logic. His 'What the Tortoise Said to Achilles' (*Mind*, n.s. iv, 1895) was very obviously an intentional logic lesson couched in a joke. Yet there is a difference between the lessons in *Alice* and the one in *Mind*. The *Alice* books were intended for children (it is only our happy fortune that they are best appreciated by adults). Children are taught lessons in order to impart new knowledge to them. Presumably, the child, given that he pays close attention, etc. can learn many things about numbers, words, and arguments – as well as cabbages and kings – from reading *Alice*. But the *Mind* article was not written for children. Only very educated persons have ever read it in a serious tone. While lessons for the novice are meant to provide new knowledge for him, lessons for the expert can only be intended to correct aberrations in his understanding.

Now the lesson of the *Mind* paper is clear: logical rules and argument premises should always be distinguished from one another. Since this is nothing new – not even in 1895 – Carroll must have believed that some logicians had ignored this distinction and thereby risked the dangers of logical confusion. But who? Who was the Tortoise?

The Tortoise held a very odd view of deductive arguments, which actually renders the process of deduction impossible. In a deductive argument a conclusion is drawn from what is known, premises. A statement is a premise whenever acceptance of its truth is required for drawing a given conclusion. Suppose we deduce C from A and B, claiming that *our acceptance of the truth of A and B is sufficient to establish our acceptance of C*. Now, in order to conclude C we had to accept A and B, but in addition we had to accept that our acceptance of A and of B was sufficient to establish our acceptance of C (i.e. we had to accept the truth of the italicized claim above). If we call this claim (that A and B are sufficient for C) D, then in order to draw C from A and B we must first be able to draw C from A, B and D. But, of course, this process will continue *ad infinitum*, so that in order to draw any conclusion from any set of premises we must first engage in an infinite series of other deductions. Thus, deduction is impossible.

Carroll's argument here is a *reductio ad absurdum* (his favourite kind). Since deduction is in fact possible, one of the Tortoise's assumptions must be rejected. Of course, the assumption to be rejected is that rules of inference can be assimilated to premises. The effect of such an assimilation is that we have a surfeit of premises but not rules of inference. Now a rule of inference *is* something we must accept in order to draw conclusion. But it is not the case that whatever must be accepted in order to draw a conclusion is a premise.

There has been, since *Principia Mathematica*, a debate among logicians about the status of, and distinctions among, rules of inference, axioms, theorems, etc. But it is chronologically impossible for the Tortoise to have represented any side of any of these issues. Moreover, no historical survey reveals any logician who has ever believed, as the Tortoise does, that rules of inference are merely hidden premises. The Tortoise seems to represent no logician. But what, then, was the point of Carroll's lesson? If his lesson was not for the uninitiated and no expert has committed the error which is exhibited, what was the point of Carroll's joke?

The logical mistake of the Tortoise is obvious. No one doubts that the Tortoise was a bad logician (or at least, a bad philosopher of logic). Even the beginning logic student can eventually pick-out the logical confusion which plagued the Tortoise. Yet that same confusion need not manifest itself in such an obviously absurd way. The Tortoise failed to appreciate the important distinction between rules of inference and argument premises. He therefore allowed rules to act as premises. But another logician, plagued by the same confusion (viz. failing to distinguish between rules of inference and arguments premises) might act differently. He might assimilate premises to rules of inference (rather than rules to premises).

Let us pretend that the Turtle was just such a logician. He would argue against deduction like this: Any deduction such as

All men are mortal

Socrates is a man

Therefore, Socrates is mortal

must beg the question. In order to accept the truth of the premises we must already accept the conclusion. We cannot know that all men are mortal unless we already know that Socrates is mortal (and that, at least, a large number of other particular men are mortal). So statements like' All men are mortal' cannot occur as premises; for if they could they would render such arguments circular. Such statements then are not premises at all – they are rules. In the case above, 'All men are mortal' is a rule which allows us to draw 'Socrates is mortal' from the single premise 'Socrates is a man'. Such a derivation is patently not deductive – it is inductive. Deduction is disguised induction.

It should be obvious now that the Turtle was J.S. Mill. Carroll knew Mill's influential and widely-read *A System of Logic* (London, 1872) well. Mill was his century's premier empiricist logician in England. Characteristically, Carroll did not attack Mill's view of deduction directly. Carroll said, in effect, that the failure to distinguish rules of inference and premises of arguments is a mistake. It results in one of two equally possible further errors: assimilating rules to premises or assimilating premises to rules. The first of these alternatives leads to such obvious absurdity that the initial mistake which fathers both it and the second alternative must be intentionally avoided. One might just as well assimilate rules to premises as premises to rules. Because both mistakes have the same origin they are equally absurd, though one may appear more so than the other (recall the Mad Hatter's remark, "Why you might just as well say that 'I see what I eat' is the same as 'I eat what I see'!"). Carroll poked fun at the logical theory of the Tortoise, knowing that while a tortoise is not a turtle, they are evolutionary brothers.

Note 2021:

I wrote three pieces on Lewis Carroll's 1895 article in *Mind* ("What the Tortoise Said to Achilles"), a Carrollian puzzle that has been exercising logicians for more than a century and a quarter now. My three attempts to say something about what the Tortoise said began with this one (1973), the next one two decades later (1993), and the latest an additional two decades later (2016). One would think that the excessive time required to revisit the Tortoise would have guaranteed that I made significant increase in my understanding of Tortoise logic. I doubt there is much evidence of that (though there is evidence that I had yet to grasp the importance of allowing the use of non-masculine pronouns in my texts).

In this first attempt, armed with the confidence of youth, aided by the audacity of believing that a couple of encounters with the Tortoise would suffice, I aimed to answer the question: Who is the Tortoise? What actual logician, if any, held the view of logical inference taken by the Tortoise, who assimilated rules of inference to premises? No human logician held such a view. However, J.S. Mill did claim that premises could be assimilated to rules of inference – Turtle logic. Carroll's mocking of the Tortoise was an indirect attack on Mill, the Turtle.

2 Shoelaces Used by Lewis Carroll

Jabberwocky, (23), 1975, p. 81

Dear Sir,

For many years now too much of Carrollian scholarship has been directed toward revealing, explicating, or exploring such trivial matters as the possibility of a Dodgson/Carroll split psyche, the philosophical puzzles surrounding the tea-party, the linguistic theories of Humpty Dumpty, or the logic of Jabberwocky. Finally, however, the tide is turning. The true Carrollian has now been treated to a superb discussion of an important and interesting (though, until now, unappreciated) aspect of Carroll's work. Warren Weaver, in *Jabberwocky*, Winter 1975, has finally foregone psychology, history, theology, philosophy, logic, etc. and has gotten down to what we have always yearned to know about that strange creator of Alice: colours of ink used by Lewis Carroll.

Weaver's boldness has been infectious. As a result I here record the findings of my own researches into another ignored yet important and interesting feature of Carroll's life and works.

Roughly speaking, Carroll used black (often faded to greyish) shoelaces until March 10, 1868; then brown until the spring of 1888; and finally blue until his death, whenever that was.

These dates are firmly fixed by references in the Diaries. There are but a few exceptions to the schedule outlined above. For example, on April 22, 1869, Carroll wore blue shoelaces and on January 1 and March 6 of 1874 he wore one blue and one brown. The political and theological ramifications of this are obvious and require no further comment. Also, on March 13, 1891, Carroll began a week of wearing no shoelaces at all. The following week he reversed his position, wearing laces but no shoes. There was a mixed period in late 1893 recorded in the Diaries as follows:

Blue	on January 1, 2, 3
Blue	on March 6, 7, 8, 9, 10, 13
Blue	on April 22, 23
Black	on July 4, 1862

Curiously, Carroll not only went through several important changes in shoelaces colour, but matched these changes with appropriate alterations in the colours of his shoes! But that matter is too lengthy to discuss here and so the results of my researches there must await another report.

Yours etc.

Note 2021

There has been a serious examination of the various colours of ink that Carroll used at different times as an aid in determining the dates of some of his manuscripts. I thought a light hearted parody would be in order. After this, it would be difficult for anyone to believe that I ever hoped to be a serious scholar of Lewis Carroll's work (especially his work on logic).

3 Lewis Carroll and the Logic of Negation

with Nora Gilday

Jabberwocky, (5), 1976, pp. 42-45

> "Contrariwise," continued Tweedledee, "if it was so, it might be; and if it were so, it would be; but as it isn't, it ain't. That's logic."

It has been difficult for scholars to make a fair judgment concerning Carroll as a logician. Either his logical work is ignored altogether in favour of his fiction, or it is fused and confused with his mathematical work. Carroll is generally considered to have been a fairly mediocre mathematician – clever in certain trivial ways, but not exceptionally competent over-all. But Carroll's logic is not mathematics. If he was not as good a logician as a story teller, was he at least a better logician than mathematician? We propose in this brief note to make a beginning at that by investigating one small but important aspect of Carroll's logic – his theory of negation.

Let us begin with a few remarks on Carroll's place in the history of logic. In the last quarter of the Nineteenth Century logic was in turmoil. The transition from the old traditional logic to the new mathematical logic was just beginning. G. Frege's ground-breaking *Begriffsschrift* was published in 1879 – though it was largely unknown for some time. In England, while Whately's *Elements of Logic* was still being read, so were Boole, De Morgan, and Jevons. But here seems to be no evidence that Carroll was part of, or even very interested in, the new logic. He remained, in keeping with his character, a logical traditionalist. Nevertheless, a close study of his *Symbolic Logic* reveals the germs, at least, of several ideas which have come to hold central places in the debates on logic which continue today. There is a very great difference between the traditional theory of logical negation and the contemporary theory of logical negation. Where does Carroll's theory stand with respect to these two? We believe that what little Carroll says about negation in *Symbolic Logic* reveals a fairly well conceived notion of negation, which, while flawed in a fundamental way, is nonetheless ripe with a logical understanding not always found in today's logic. Logic is, to put it very simply, the study of argumentation – in particular, the formal grounds for inferring one sentence from a set of other sentences. *Formal* logic claims that whether or not a sentence (conclusion) can be inferred (follows) from a set of sentences (premises) is a matter of the logical forms of the premises and conclusion. Both traditional and contemporary logicians are formalists in this sense. What they disagree about is how to determine the logical forms of sentences. Consider the sentence 'Alice is not a flower'. The contemporary logician says that the sentence can be analyzed into a simple sentence, 'Alice is a flower' and a negation 'not'; and that the only thing important here is the 'not'. The simple sentence can remain unanalyzed. It is thus replaced by a "sentential variable", p. The term 'not' is usually symbolized by '~', so that the whole sentence is formalized (symbolized) as '~p'. The traditionalist, however, holds that all sentences must be analyzed further into their subjects and predicates. Both the contemporary and traditional logicians minimize the dangers of letting formalized sentences contain nonformal parts by replacing these parts with variables (marks or symbols that can stand for different things in different cases, like the x's and y's of algebra). But, while the contemporary logician takes entire sentences to be in need of such replacement (thus 'p' replaces the entire sentence 'Alice is a flower'), the traditional logician takes only the subject and predicate of such sentences to be so replaceable. He would formalize 'Alice is not a blower' by 'A is not F'. The contemporary logic we are speaking of here is called "sentential" because its variables are sentences. The traditional logic is called "term logic" because its variables are terms.

Obviously, what a logician thinks about the nonvariable (formal) elements of sentences will be strongly determined by what he takes as variables. Logicians recognized a variety of formal elements. In English they are found in words such as 'all', 'some', 'are', 'non', 'not', 'unless', 'no', 'if', 'and', 'or' and so forth. Some of these formal elements when added to sentential variables form new sentences. Thus 'and' when added in an appropriate way to 'Rose is a flower' and 'Alice is not a flower' results in a new, more complex sentence, 'Rose is a flower and Alice is not a flower'. Other formal elements when added to terms result in sentences. Thus adding 'are' to the terms 'men' and 'animals' results in the sentence 'Men are animals'.

Negation is a process which results in a new sentence. In English, negation is achieved by words like 'non', 'non', 'un', 'it is not the case that', 'it is false that', etc. The logician can simply take All of these to be variations on 'not'. Now we can introduce 'non" into a sentence in more than one place. We can say (1) 'Alice is unhappy' and (2) 'Alice is not happy' and (3) 'It is not the case that Alice is happy'. Here is where traditional and contemporary logicians disagree about negation. The traditionalist says that in any subject-predicate sentence the predicate can be either affirmed or denied of the subject. A predicate is denied of a subject by negating the copula 'is' or 'are'). Negating a copula is not the same as negating a predicate. 'S is P' affirms P of S, 'S is not-P' affirms not-P, the negation of the predicate P, of S. 'It is not the case that' is a long-winded way of denying a predicate (i.e., negating a copula). So, for a traditional logician, sentences (1) and (2) are different. (1) affirms a negative predicate while (2) denies a positive predicate), and (3) is logically equivalent to (2). The contemporary logician claims, however, that all three of our sentences are logically equivalent. They are all sentential negations of the simple sentence 'Alice is happy', which is in no further need of analysis. (3) is the standard form of all three.

Let us call the 'not' of 'S is not-P' *predicate negation*, the 'not' of 'S is-not P' *copula negation* and the '~' of '~p' *sentential negation*. Which of these kinds of negation did Lewis Carroll recognize? As we would expect, there is no admission of sentential negation anywhere in *Symbolic Logic*. He did accept the traditional view of predicate negation. He even recognized the Aristotelian notion, often obscure in other traditionalist works, that a sentence like 'No S is P' is logically equivalent to a sentence containing a predicate negation (viz. 'All S is not-P'). What is unique and interesting in Carroll's theory, however, is the fact that he rejects copula negation altogether. Thus Carroll was not a strict traditionalist. In the old logic it was copula more than predicate negation which was taken to be fundamental. Carroll claimed in section 3 of the Appendix to *Symbolic Logic* that the question of which kind of negation is involved in a sentence such as Alice is not a Flower' is "merely a matter of *taste*, since the two forms mean exactly the same thing."[11] What suited Carroll's taste was to take all sentences as affirmations, construing denials (sentences with copula negations) as affirmations of negative predicates (sentences with predicate negations). He said that if we form a "Dichotomy by Contradiction" (i.e., two predicates P and not-P) it is better to say of a subject "either that it 'is' in one, or that it 'is' in the other." What he meant was that in any subject-predicate sentence it is best to take the copula as unnegated and construe the predicate as either negated or

---

1 See *Symbolic Logic and The Game of Logic*: Dover Publications, New York 1958, p. 167.

unnegated as the case may be. For Carroll, predicates could be negated but not denied. What might look like a denial ('S is-not P') was taken as the affirmation of a negative predicate ('S is not-P').

Carroll was simply wrong in saying that the issue here is just a matter of taste. While he justly deplores the logical habit of his day of ignoring negative predicates, his rejection of predicate denial (copula negation) would have far-reaching and logically disastrous consequences. That he was unaware of these consequences can only be due to the fact that his interest in logic was decidedly nontheoretical. Had Carroll tried to develop a fuller logical theory he would have quickly recognized the need for copula negation in order to account for a variety of logically important distinctions. Even if he had no interest in producing such a theory, a careful reading of Aristotle would have been sufficient warning to him.

So Carroll was not a theoretical logician. The weakness of the theoretical base beneath his notion of negation is sufficient evidence of this. Nor was he a practical logician. For Carroll logic was a play, mental exercise. What other logician would write *The Game of Logic*? Nevertheless, Carroll was aware of theoretical issues, especially those involving paradox. At the very least we can say that almost a century ago Lewis Carroll did what logicians only now are beginning to do again – think about the hidden complexities of the simple 'not'. A fair judgment of Carroll's logic in general awaits judgments of those parts which are unique or characteristic of his logic. This in turn demands a careful examination of each of those elements. Negation is but one of these. Carroll also held interesting views on the nature of deduction, existential import, the quantification of singular terms, and the extension of syllogistic. Study of these topics in Carroll's logical work should determine whether or not he must remain what he once called himself, "an obscure Writer on Logic, towards the end of the Nineteenth Century."

Note 2021

The notion of what negation is and how it operates in logic was fairly well understood by most logicians until the end of the 19th Century. Logical negation only became a matter of debate when the new Mathematical Logic, initiated by G. Frege, determined that there could be but one kind of negation – *sentential* negation. Only entire statements can be negated. As a logician,

Carroll was a traditionalist. In his *Symbolic Logic*, he took a bold stand on the nature of negation, arguing that from a logical point of view there is no sentential negation. Yet, in contrast to most traditionalists, Carroll claimed that *copular* negation (as in 'Alice is-not a Flower') is equivalent to *predicate* negation ('Alice is not-a Flower'). It turns out that there is much more to logical negation than Carroll realized. Nonetheless, his treatment here reveals that Carroll ought not be considered as simply "an obscure Writer on Logic."

4 A Note on the Sense of *Jabberwocky*

Jabberwocky, (5), 1976, pp. 93-94

There is more than one way to produce linguistic nonsense. Lewis Carroll knew them all.

One easy way to produce nonsense is simply to ignore the normal rules of grammar. Thus, 'only or John if it not run' is a piece of nonsense although each word in the sequence is perfectly sensible. A second, slightly more subtle kind of nonsense results from rejecting rules of logic which prohibit contradiction. Thus 'All roses are red, but some are white' and 'The committee has four members and does not have four members' are examples of logical nonsense. A still more subtle kind of nonsense results from ignoring certain meaning restrictions on the kinds of terms that can be predicated of certain kinds of subjects. For example, 'The number two is red' and 'Loyalty drinks only white wine' are ordinarily quite nonsensical. *Jabberwocky*, however, illustrates still another kind of nonsense. Here the rules of grammar and logic are flawlessly adhered to. Yet the result is sentences which seem to have a kind of sense but are nonetheless nonsense. As Alice says, 'It seems very pretty, but it's *rather* hard to understand! Somehow it seems to fill my head with ideas only I don't know exactly what they are!'

A sentence is more than just a list of words. The words of a sentence can be used to say something sensible only when they are joined together in a certain number of very limited ways. The rules governing how words are so joined are the rules of grammar. Each language has its own unique way of joining words to make sentences, i.e., its own grammar. Whatever the grammar might be, words are joined by the use of a very few special phrases, words and word parts, which linguists call *morphemes*. A sentence is made up of 'content' words plus morphemes, in accordance with grammatical rules governing the order of each element in the sentence. Take away the content words of a sentence and what remains is the grammatical form. Such forms can be filled in many ways. Thus, the form '___'s ___ is ___ ' might be 'John's hat is green' or 'Mary's mother is ill'. Again, '___ed a ___y ___ in the ___s' might be completed as 'He spied a happy hermit in the woods' or as 'Fido bit a naughty boy in the legs'. Some common and important English morphemes are: ' 's' which attaches to singular nouns, pronouns or noun phrases to make them possessive; 'ed' which attaches to verbs to make them past tense; and 'ly' which attaches to adjectives to make them adverbs. There are many more. Needless to say, the morphemes which constitute the form of a sentence play a crucial role in giving sense to that sentence. 'John hat green' makes some sense, but not much. What makes Alice feel that the poem has some sense is the invariably correct grammar of each of its sentences. On the other hand, the content words almost always have either no sense, or unclear senses, or vague senses, or ambiguous senses. In *Jabbewocky* sentences, almost all the burden of sense is carried by form rather than content.

Here is the morphemic structure of the first verse.

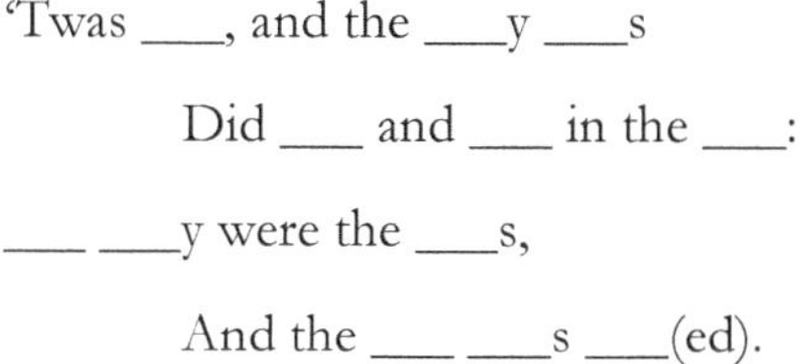

'Twas ___, and the ___y ___s
Did ___ and ___ in the ___:
___ ___y were the ___s,
And the ___ ___s ___(ed).

In writing *Jabberwocky* Carroll seems to have had a keen clear picture of the morpheme structure of the English language. To appreciate this it should be kept in mind that the serious, systematic science of linguistics did not begin until after the second world war.

Note 2021

Carroll was a master of the art of creating nonsense with words. There are many ways of doing this. One that he was particularly successful with was the production of sentences that were both perfectly *grammatically* correct and *logically* correct, seemed on first sight to mean something – but were quite senseless, pure nonsense. My note offered a simple example (a verse from *Jabberwocky*) of how Carroll's sentences could be deconstructed to reveal their grammatical and logical correctness, while still lacking any real sense.

5 Lewis Carroll on Logical Quantity

Jabberwocky, (12), 1983, pp. 39-41

In this note I want to discuss briefly two different features of Lewis Carroll's theory of logical quantification. The first is a notion shared by most pre-Fregean logicians. It is a mistaken one. And, in spite of this being demonstrably the case, there are some today who still entertain it. The second is an idea which is, I believe, original to Carroll, one which, had it been formulated more clearly and fully, might have come closer to the truth.

1. Peter Geach has argued (*Reference and Generality*, Ithaca 1962, pp. 11ff, and *Logic Matters*, Oxford 1972, pp. 54-59) that quantified expressions are not referential. In doing so he makes two points: (i) that viewing quantified expressions as referring phrases is a result of taking categorical sentences as class sentences; and (ii) that 'no x' is a quantified expression along with 'every x' and 'some x'. Geach argues, in effect, that since 'no x' does not refer, no quantified phrase refers. It can be shown that such an argument will not work (see Fred Sommers, *The Logic of Natural Language*, Oxford 1982, pp. 337-338). For while 'no x' does indeed fail to be a referring phrase, it, unlike 'every x' and 'some x', is *not* a quantified expression. 'No' is no quantifier.

Carroll took 'no' to be a quantifier. So did most logicians of his day. In *Symbolic Logic* (W.W. Bartley, ed., New York 1977) Carroll says:

> A Proposition, in normal form, consists of four parts, viz. –
>
> (1) The word "some," or "no," or "all."
> This word, which tells us *how many* Members of the Subject are also Members of the Subject are also Members of the Predicate, is called the 'Sign' of Quantity.') (p. 68, cf. 69)

It is clear that if one takes standard categorical sentences to be class sentences (e.g., reading 'some A are B' as 'some members of the class of A-things are members of the class of B-things') then one is led to treat quantifiers as applicatives which indicate *how many* members of a given class are being referred to. On such an interpretation it is then easy to view 'some' as indicating 'one or more' and 'no' as indicating 'zero'. Contra Carroll, however, 'no x' is no referring expression referring to 'no x' (nor to the null set). But, contra Geach, this is because 'no', unlike 'every' and 'some' is no quantifier at all.

What is 'no'? It is a portmanteau word. What it says is: 'not a' (or 'not some', or 'not any'). In effect, a sentence of the form 'no x is y' can be logically regimented as 'not an x is y', the contradictory of 'some/an x is y'. Consider 'No creature was stirring'. This denies (i.e., is the contradictory of) the sentence 'A creature was stirring', thus it is equivalent to 'Not a creature was stirring'. It is unfortunate that Carroll failed to fully understand the logical nature of 'no', taking it as a quantifier rather than the composite sign of particular quantity and sentential denial. But most of his contemporaries, as well as some of ours, made the same mistake.

(2) Modern mathematical logicians equate universal affirmations (A-form categorical) with their obverses. Thus 'every A is B' is taken to be equivalent to 'no A is not B' (or 'not an A is B').

Symbolically:

$$(\forall x)(Ax \supset Bx) \equiv \sim(\exists x)(Ax \;\&\; \sim Bx)$$

Nonetheless, there appear to be counterexamples to such an equivalence. Consider: 'No unicorn is unclean' and 'No horse is prime'. These are arguably true. But 'Every unicorn is clean' and Every horse is nonprime' are surely false (though for different reasons).

Modern logicians also treat conditional (hypothetical) sentences as material implications (cf. Bartley's discussion at pp. 447- 448). Thus they are true whenever their antecedents are false. 'Every A is B' is read as 'Everything is such that if it is A then it is B'. The conditional 'if it is A then it is B' will thus be true when 'it is A' is false, which will be the case whenever there are no A-things. Since there are no unicorns 'Every unicorn is clean' is taken to be true. In fact, on this theory, whatever one says about all unicorns will be true! Now traditional syllogistic logicians, including Carroll, accepted obversion. But they avoided calling 'Every unicorn is clean' true by demanding subalternation. According to subalternation 'every A is B' logically entails 'some A is B' When 'some A is B' is false (as with 'Some unicorn is clean') so is 'every A is B'. In *Symbolic Logic*, however, Carroll offered two slightly different accounts of obversion and subalternation.

According to his first account (pp. 74-75) 'every A is B' is *equivalent* to a conjunction of 'some A is B' and 'no A is not B'. Thus he calls universally quantified sentences 'Double Propositions'. Now this theory nicely accounts for subalternation (since a conjunction entails any of its conjuncts). But it voids obversion (since the traditional obverse of a universal is no longer equivalent to it but only to part of it).

Carroll later (p. 121) adopts a slightly different thesis. He still takes universals as 'Double Propositions', and while one conjunct remains the classical obverse, the other conjunct is now not the subaltern but what might be called the existential import. According to this thesis 'every A is B' is logically parsed as 'some A exists' and 'no A is not B'. As before, classical obversion is blocked; but now so is subalternation (since 'some A exists and no A is not B' entails 'some B exist' but not 'some A is B'). The new formulation is meant, according to Carroll, to eliminate 'superfluous information'.

The various positions mentioned here can easily be summarized using 'A' for the sentence 'every A is B', 'I' for 'some A is B', 'O' for 'some A is not B', 'a' for 'some A exists', '~' for denial, '&' for conjunction, '⊢' for logical entailment, and '≡' for logical equivalence. The modern logician holds:

(M1) $A \equiv \sim O$ (obversion)

(M2) $\sim(A \vdash I)$

Carroll's first position was:

(C1) $A \equiv (I \,\&\, \sim O)$

(C2) $A \vdash I$ (subalternation)

His second position was:

(C1a) $A \equiv (a \,\&\, \sim O)$

(C2a) $A \vdash a$ (existential import)

As a logician Carroll was a man of his times. Like Boole, De Morgan, Keynes, Venn, Jevons, etc., he helped formalize the traditional syllogistic. His particular contributions come from his method of diagrams, his clearer concept of deduction, and his insistence on paying attention to the puzzle cases instead of the easy ones. Nonetheless, he shared with his contemporaries (and some moderns) confused and sometimes quite mistaken notions about logical quantity.

Note 2021

While Carroll did pay attention to the nature of logical quantity in his *Symbolic Logic*, he nonetheless held certain notions about this. These came to be shared by some mathematical logician in the next century (e.g. P. Geach). In particular, Carroll mistakenly took phrases of the form 'no x' as quantifier expressions. They are not. He also struggled to account for inferences that seem to require the application of three traditional requirements: the rule of *obversion*, the rule of *subalternation*, and the retention of *existential import*. Carroll was plagued here by his ambivalence about the proper logical analysis of universal affirmations.

6 Carrollian Things

Jabberwocky, (13), 1984, pp. 96-100

Not long ago I reread E. Coumet's '*Jeu de logique, Jeux d'univers*',[1] which naturally led me to return to the logic of DeMogan and Venn. For Lewis Carroll, qua logician, belonged to that company of Nineteenth Century British logicians which was to be so suddenly (and rudely) repudiated by the Fregeans who came to dominate logic in our century. While accepting the power, and even the beauty, of modern mathematical logic, I nonetheless refuse to ignore the many keen insights of men like De Morgan, Venn and Carroll. Indeed, at least one contemporary logician, the American philosopher F. Sommers, has argued recently for the ascendency of their brand of logic over today's.[2]

Traditional logic was a logic of natural language. Contemporary logic has its origin in the search for the foundations of mathematics. The great importance and nobility of mathematics notwithstanding, it is the medium of natural language in which the vast majority of inferences are carried out. Concomitant with his rejection of natural language as an adequate medium for logical reckoning, the mathematical logician rejected the traditional logician's views of negation, quantity, and meaning. Moreover (and this is the topic I wish to pursue here), modern logicians have imported into logic the notion of a bare particular – a Thing. According to today's standard logic every thing exists. A thing is whatever is taken to be an object of reference (where the burden of reference, as W.V.O. Quine has so often said, is squarely on the shoulders of the bound variable, or pronoun). According to this view, things are bare particulars having no attributes. For Quine has admitted:

The pronoun is the tenable linguistic counter part of the untenable old metaphysical notion of a bare particular.[3]

The variable is the legitimate latter-day embodiment of the incoherent old idea of a bare particular.[4]

Quine never makes clear just *what* is 'untenable' or 'incoherent' in the idea of a bare particular. Perhaps it is just that the idea is old. He seems to like it well enough in its modern dress.

Carrollian logicians took the notion of a bare particular, a thing without attributes, a Thing, as untenable and incoherent in *any* guise. In the

'Preface to Fourth Edition' of *Symbolic Logic* Carroll referred to the 'bewildering question "What is a Thing?"'[5] And in *The Game of Logic* he referred to the question 'Can a Thing exist without any Attributes belonging to it?' as a puzzling question' not worth answering. He recommended that we 'turn up our noses, and treat it with contemptuous silence, as if it really wasn't worth noticing.'[6] How is it that traditional logicians (like Carroll) and their supporters (like Sommers) avoid bare particulars – Things, while modern logicians (like Quine) do not? I believe logicians are seduced into accepting Things by their misunderstanding of term negation. Further, the appropriate safeguard against Things is an understanding of how reference in ordinary uses of natural language is restricted, especially by domains of discourse.

While contemporary logicians recognize only one kind of logical negation (sentential), the traditional logician recognized at least two (denial and term negation).[7] Most importantly, the contemporary logician does not accept term negation as anything more than a stylistic variant of sentence negation. In spite of the parsimony achieved by this modern view, one cannot gainsay the abundance of negated terms in use in natural language. And these uses are not equivalent to corresponding uses of negated sentences. Consider the fact that 'Socrates was not a Maosit' does not logically entail 'Socrates was a nonMaoist'. Modern logicians take all negators (e.g., 'not', 'non') as applying to entire sentences. Thus, they parse both of these sentences as 'Not: Socrates was a Maoist'. (Ironically, Carroll seemed to make just the opposite kind of mistake, arguing, in effect, that negation is always term rather than sentential.[8]) Now a term and its negation (say 'Maoist'/'nonMaoist') determine two exclusive sets. We are fairly certain about what belongs to the first, but the population of the second is far from clear. Surely people who are Nazis, or Social Democrats, or British Conservatives belong to the set of nonMaoists. But what about my dog, Socrates, the number 2, ... ? Let the set of nonMaoists (the counterset of Maoists) contain absolutely anything we might refer to which is not Maoist. The counterset will indeed contain both Mrs. Thatcher and my dog and the number 2. And, while we know what attribute Maoists all have in common, by virtue of which they are Maoists, we have no idea what attribute all such nonMaoists share. We cannot even say that all nonMaoists are persons, or even material objects. In fact, the most we *can* say is that all nonMaoists are Things.

In our ordinary use of natural language we rarely (probably never) refer to Things. We do of course make frequent reference to things. The difference here is that while Things have no attributes, things do. In fact, things are always of this or that type (or sort, or category). Every thing is a thing of some type.[9]

The importance of belonging to a type is that such membership always entails the possession of a specifiable range of attributes. Thus, while Things have no attributes, things do. Person-type things, persons, are living, rational, animal, mammalian, etc. Star-type things, stars, are inert, extraterrestrial, etc. So in ordinary discourse we make reference to things of this or that type. And the negation of a term is taken to determine not the counterset of that term but on the subset of that counterset which intersects with the type determined by the term. For example, while 'nonMaoist' determines a set (the counterset of 'Maoist') which includes all Things other than Maoists (thus including Mrs. Thatcher, my dog and the number 2), as used in ordinary discourse it determines only that set of Things which are also of the type determined by 'Maoist'. Presumably this type is simply persons. Thus, in its ordinary use, 'nonMaoist' determines not a set of Things but a set of Things which are persons, i.e., person-type things, i.e., persons (exclusive of Maoists). It follows that in its ordinary use 'nonMaoist' could determine a set containing Mrs. Thatcher, but neither my dog nor the number 2.

The idea that the denotation of a term is restricted by type considerations was not clearly recognized by Nineteenth Century logicians. But they did recognize (and far more clearly than today's logicians) a second restriction of term denotations. Every sentence when in use is used relative to some specifiable domain of discourse. What the domain is for any sentence is fairly arbitrary, subject only to the restriction that it be specified if necessary and that both the speaker and audience understand the sentence to be being used relative to the same domain. A domain can be any totality of objects – the actual world, a possible world, the world of *Hamlet*, Wonderland, etc. (any set is a possible domain). Used relative to the actual world 'Some caterpillar smokes a hookah' is false. But, relative to Wonderland, it is true. The difference here is due to a difference in the denotation of 'caterpillar'. Let us say that, outside of its use in any sentence, every term has an *extension*. All caterpillars, real, possible, fictitious, dreamed, imagined, or whatever, are in the extension of 'caterpillar'. But, every term, when used, is used in a sentence in use. And every sentence in use is used relative to some domain. The effect of using sentences relative to domains is this: the domain adds a restriction to the extensions of the terms used in the sentence. While 'caterpillar' extends over *all* caterpillars, its use in a sentence used relative to the actual world places specifiable limits on its extension. Let us call the part of a term's extension restricted by a domain the term's *denotation*. What a term denotes is determined then by its extension *and* the domain relative to its use. Using the sentence 'Some caterpillar smokes a hookah' relative to the actual world means that the denotation of 'caterpillar' is not its entire extension, but just that part of its extension which is in the domain (i.e., real caterpillars). Using the same sentence relative to Wonderland means that the denotation of 'caterpillar' is confined to caterpillars in Wonderland. While in each case 'caterpillar' extends over both the insects crawling on the elm in my garden and the mystifying character met by Alice, only the first are denoted by 'caterpillar' in the first use of the sentence (the one relative to the actual world), and only the hookah smoker is denoted by 'caterpillar' in the second use.[10]

Suppose I say, as Alice did at the Mad Tea Party, 'Nobody asked *your* opinion!' Well, *who* didn't ask your opinion? It all depends, of course, on the domain of discourse. It might be absolutely everybody (every person) who failed to ask you your opinion. But more likely, I, like Alice, use my sentence relative to some domain which restricts the extension of 'body'. When she says to the Hatter, 'Nobody asked *your* opinion', her domain is certainly not the actual world (she didn't have in mind Queen Victoria as among those who didn't ask for the Hatter's opinion); nor was her domain Wonderland (for she clearly did not even have in mind the Duchess). Her domain was just the members of the tea party. Her 'body' denoted Alice, the March Hare, the Hatter, and the Dormouse – nobody else.

Domains can be as big or small as we wish. They can be as ordinary or as extraordinary as we wish. But they cannot be ignored in accounting for the ordinary uses of terms. The modern logician parses a sentence like 'Some caterpillar smokes a hookah' as 'Something (= some Thing) is both a caterpillar and a hookah smoker'. On this analysis all we ever talk about are Things, bare particulars, But unless the modern logician is prepared to answer Carroll's questions about Things, this analysis is just plain nonsense. The traditional logician, by not ignoring the ordinary uses of natural language sentences, knew that we talk about all types of things – peoples, dogs, numbers, and caterpillars. What we do *not* ordinarily talk about are Things.[11]

Notes

1 In *Lewis Carroll*, H. Parisot, ed.: Editions de l'Herne, Paris 1971. The English translation made by P. Heath appeared as 'The Game of Logic and a Game of Universes' in *Lewis Carroll Observed*, E. Guiliano, ed.: Clarkson N. Potter, New York 1976.

2 F. Sommers, *The Logic of Natural Language*: Clarendon Press, Oxford 1982. I have discussed Sommers' logical counter-revolution in many places. See especially G. Englebretsen, *Three Logicians*: Van Gorcum, Assen 1981.

3 'The Variable and its Place in Reference,' *Philosophical Subjects*. Z. van Straaten, ed.: D. Clarendon Press, Oxford1980, p. 165.

4 'Grammar, Truth, and Logic,' *Philosophy and Grammar*, S. Kanger and S. Ohman, eds.: Reidel, Dordrecht 1981, p. 25.

5 Dover, New York 1958, p. xii.

6 Dover, New York 1958, p. 2.

7 I have discussed these differences in detail in *Logical Negation*: Van Gorcum, Assen 1981. See also G. Englebretsen and N. Gilday, 'Lewis Carroll and the Logic of Negation,' *Jabberwocky*, Spring 1976.

8 Cf. *Symbolic Logic*, p. 172. There he suggests that negators should be attached not to copulae (rendering negated sentences) but to predicates (rendering negated terms). This position was not typical among traditional logicians.

9 Recognition of this requirement is at least as old as Aristotle. See the interesting discussion of it in M. Durrant, 'Numerical Identity,' *Mind*, 82 (1973).

10 A fuller account of the notion of domains found in the logic of De Morgan, Venn and Carroll is offered in Coumet's article (see note 1).

11 I have argued that neither traditional nor modern logicians have paid sufficient attention to the logical restrictions imposed by sorting things into types. But, unlike the traditionalists, the contemporary logician does not even recognize the restrictions imposed by domains of discourse. Today's logician takes all genuine reference to be to unsorted, bare particulars (Things), which *exist*. I have yet to raise the interesting question of what existential commitments are made by the logic of De Morgan, Venn and Carroll.

Note 2021

The accounts of logical quantification, reverence, and even negation at the heart of today's standard mathematical logic seem to force logicians to accept the notion of a *bare particular*, a *Thing* devoid of any attributes. The burden of reference, then, is restricted to bound variables (anaphoric pronouns) and names. Traditional logicians like Carroll restricted their focus to ordinary, natural language. It is a language in which one speaks of all sorts of things (dogs and people, stars and numbers, not to mention ships and shoes and sealing wax or cabbages and kings), but not of *Things*. In his logical works, Carroll found the question of what is a Thing "bewildering" and denied that anything could fail to have any attributes at all (i.e., be bare). In this contribution, I briefly outline how the new logic, ignoring the crucial role played by *domains of discourse* in determining how our terms can be used to refer to things, has led to confusions among mathematical logicians about negation, quantification, and reference. Carroll and his contemporaries recognized how to avoid such confusions.

7 A Note on Carroll, Wittgenstein and Nonsense

Jabberwocky, (15), 1986, pp. 19-20

There is a commonly held view among those analytic philosophers who have taken a close and serious look at Lewis Carroll that he had the kind of interest in linguistic nonsense displayed in the work of Ludwig Wittgenstein. Prominent examples of such a comparison are those found in G. Pitcher's 'Wittgenstein, Nonsense and Lewis Carroll' (*Massachusetts Review*, 6, 1965), W. Shibles' *Wittgenstein, Language and Philosophy* (Dubuque, Iowa, 1969) and P. Heath's *The Philosopher's Alice* (New York, 1974). Now it is certainly true both Carroll and Wittgenstein displayed a deep interest in the phenomenon of linguistic nonsense. Nonetheless, what they said about nonsense, and how they said it, were quite different.

In both his early and late works Wittgenstein was intent on laying bare the sources of nonsense. In the *Tractatus Logico-Philosophicus* (London, 1922) this meant the kind of nonsense generated in the language of philosophers by their 'failure to understand the logic of our language' (4.003). Logic, according to the early Wittgenstein, sets limits to our language/thought/world (5.6 - 5.61). Nonsense is the result of trying to go beyond those limits. Logic constrains and determines the forms of propositions and thereby limits language. But these forms – and these limits – cannot be *said.* They can only be *shown* (4.1212). Nonsense is the result of trying to say what can only be shown, trying to say what cannot be said (7). In a deeper sense, of course, Wittgenstein took even his own attempts in the *Tractatus* to draw the limits of language to be themselves pieces of nonsense (6.54).

In the posthumously published *Philosophical Investigations* (New York, 1953) Wittgenstein was still looking for the source of linguistic nonsense (and doing many other things as well). He still believed that nonsense was to be found in the language of philosophers. But he no longer believed that it was due to their failure to understand the logic of language. In his later works Wittgenstein was convinced that ordinary, non-philosophical language was 'in order', and that nonsense resulted when philosophers 'misuse' ordinary language. The nonsense of philosophers' language is generally disguised, hidden. He saw his task as that of revealing this nonsense, of making it explicit (464, 524). It is language's ability to bewitch (109) which leads philosophers to produce nonsense. The proper remedy is to be shown language working it its ordinary, every-day manner. Wittgenstein wanted to show the fly (philosopher) the way (attention to ordinary language use) out of the fly-bottle (nonsense/philosophy) (309).

It is hard to imagine a more serious, sober, often obscure approach to the subject of linguistic nonsense than Wittgenstein's. Carroll's approach was none of these. He recognized, for example, that linguistic nonsense was often generated simply by and excessive adherence to the syntactic and semantic constraints on language. He knew well how the limits placed on language by logic produced nonsense when one bumped up against them too hard. Bu this intention was never to give a careful account of such nonsense. Carroll may have been a passable logician, but he was no philosopher. What he did see was that linguistic nonsense can very often be surprising, bewildering, disconcerting, and, not infrequently, funny.

Carroll, unlike Wittgenstein, actually said very little about linguistic nonsense. Wheat he did was show it. Examples of nonsense, linguistic and non-linguistic, fill the pages of Carroll's work. But, while he may have genuinely believed that sensitivity towards the tendency of language to generate nonsense inhances one's ability to use language clearly, rationally, effectively, he never said so. Though, it must be added, in his logical works he did, in an often whimsical way, draw attention to the logical resources of language, resources which, if not properly cared for, could led to logical nonsense. In this Carroll, unlike Wittgenstein, laid the blame for nonsense on the language user, not the language.

Finally, in fairness, it must be noted that Wittgenstein, that paradigm of seriousness, did once remark that one could produce a serious work of philosophy consisting entirely of jokes. But, of course, he had in mind 'deep' jokes (*Philosophical Investigations*, III).

Note 2021

A number of analytic philosophers have seen important areas of comparison in the business of linguistic nonsense as it appears in the writings of Lewis Carroll and Ludwig Wittgenstein In this brief contribution, I tried to sketch how, early on, Wittgenstein took *philosophical* nonsense to be the result of philosophers failing to adequately understand the "logic of our language" and then, later, saw it as the result of philosophers "misuse" of our language. This was a serious, sober, even obscure take on linguistic nonsense. One must strain to see grounds for comparing it with Carroll's. For Carroll *said* little about nonsense (linguistic or otherwise) – he simply *showed* it. There are two ironies here. Wittgenstein held that what determines the limits of language can only be *shown* – not *said*; and, while Carroll revelled in the fun of encountering nonsense, it was Wittgenstein who once said that a serious work of philosophy could be written consisting entirely of jokes.

8 Are Some Pigs Pink? Geach on Carroll

Jabberwocky, (17), 1988, pp. 8-10

At the very beginning of *The Game of Logic* Lewis Carroll took care to distinguish between substantive terms, which are used in sentences to mean Things,[1] and adjective terms, which are used to mention Attributes. He then goes on to say:

> You may put 'is' or 'are' between the names of two *Things* (for example, 'some Pigs are fat Animals') or between the names of two *Attributes* (for example, 'pink is light red', and in each case it will make good sense. But, if you put 'is' or 'are' between the name of a *Thing* and the name of an *Attribute* (for example, 'some Pigs are pink'), you do *not* make good sense (for how can a Thing *be* an Attribute?) unless you have an understanding with the person to whom you are speaking. And the simplest understanding would, I think, be this – that the Substantive shall be supposed to be repeated at the end of the sentence, so that the sentence, if written out in full, would be 'some Pigs are pink (Pigs)'. And now the word 'are' makes quite good sense.[2]

Peter Geach happens to be one of those rare contemporary British logician-philosophers who is willing (and able) to pay any serious attention to the work of logicians which appeared before the Fregean revolution in logic, which began in the 1880s and continues today. To be sure, as a good Fregean, Geach tends to make only negative judgments about such logical work, but at least he doesn't entirely ignore logicians like Aristotle, Ockham, Aquinas, Leibniz, Buridan, or the Nineteenth Century British algebraic logicians. Boole and De Morgan are the giants of this last group, but Carroll belongs there as well. And Carroll, like the rest, is not spared Geach's disapproval.

Geach has never tired of attacking traditional logicians for their perceived adherence to what he calls 'the old two-name or identity theory of predication'.[3] He takes the passage quoted above as a prime example of a logician in the throes of this old theory. Now according to the two-name theory, in any true subject-predicate sentence (one made up of a substantive and an adjective connected by 'is' or 'are', to use Carroll's vocabulary) both the subject term and the predicate term name an individual object (Thing for

Carroll), and, most importantly, they name the same object. This theory is meant to contrast unfavourably with the one formulated by Frege, assumed by virtually all contemporary mathematical logicians, and defended vigorously by Geach. According to the new 'function-name' theory, in any true subject-predicate sentence only one term (the name/subject) names an individual object while the other term (the predicable/function/name) applies to or is true of the object named by the other term.

If you are not a logician you probably find all this either boring or even a bit silly. But it is the kind of question which had great import for logic in general, and logic – the careful examination of our modes of reasoning and rational expression – is as important as we think being rational is. And it is precisely the kind of question logicians like Aristotle, Carroll, Geach and myself find not only important but fascinating.

All I want to indicate in this brief note is that I believe that both Carroll and Geach have got things wrong here. As it is formulated by Geach, Carroll does subscribe to the two-name theory of predication, and that version of the theory is, indeed, mistaken.[4] Carroll, like most traditional logicians, seemed to assume that only noun phrases, (Carroll's substantives), such as nouns, proper names, or personal pronouns could name. Since predicates are often not noun phrases (they are often verb phrases, prepositional phrases or adjectives), in order for predicates, like subjects, to name something, they must first be turned into noun phrases. This is exactly what Carroll did with the predicate of 'some pigs are pink'. Here the predicate term 'pink' is an adjective – thus it does not name, according to the old theory. So Carroll replaced it with the term 'pink pigs', which is a noun phrase, a substantive, and can thus name an object, or Thing. Now while Geach, like any Fregean, rejects this theory of predication, he assumes, along with traditional logicians like Carroll, that only noun phrases, substantives, can name. He, unlike Carroll, may be unwilling to turn all predicate terms into noun phrases, preferring to leave them as they are and give them some semantic role other than naming (viz., applying to or being true of). But he still accepts the old notion that only noun phrases can name. To be fair, traditional logicians did not usually call the semantic relation between a noun phrase and the appropriate object 'naming'. They tended to use words like 'denote', 'refer', 'apply to', 'introduce', 'mention', 'stand for', 'supposit for', etc. It may be difficult for some to see 'pink' in 'some pigs are pink' as, say, standing for the colour pink, but it's even harder to see 'pink' as the name of anything (and, of course, that is just why Geach insists on using 'naming').

Suffice it to say, I do not believe that one must assume that naming, denoting, standing for, etc. objects is a role scripted only for noun phrases. No argument (good or bad) has ever been advanced for such a view. Moreover, theories such as Carroll's and Geach's, which assume it, generally tend to have to say some pretty silly things about perfectly simple ordinary sentences. Thus Carroll had to say that when we say 'some pigs are pink', without the understanding that 'pink' really means 'pink pigs', we do not make good sense! And Geach is forced to say that when we say 'some pigs are pink' the term 'pink' cannot stand for anything – not even the colour pink!

You and I know that some pigs are pink, pure and simple. Now, do pigs have wings?

Notes

1 I have discussed Carroll's notion of Thing in 'Carrollian Things', *Jabberwocky*, Autumn 1984, pp. 96-100.

2 This is from the Dover edition, 1958, pp. 2-3.

3 See, for example, *Reference and Generality*. Cornell University Press, Ithaca, 1962, pp. 34-36; and *God and the Soul*, Routledge & Kegan Paul, London 1969, pp. 43-44.

4 A version of the two-name theory can quite easily be defended (the one Geach describes cannot). See Aris Noah, 'The Two Term Theory of Predication', in G. Englebretsen, ed., *The New Syllogistic*, Peter Lang Publ., New York 1987, pp. 223-243.

Note 2021

One of the claims critics make against traditional logic is that it is committed to a "two-name" theory. The theory holds that subject-predicate statements are to be analyzed (for logical purposes) as copulated pairs of terms and those terms *name* the same thing. Standard modern logic, in contrast, construes such statements as constructed from a *function* (the predicate) applied to a name (the subject). Now Carroll did seem to subscribe to the two-name theory. He held that naming can only be done by the use of noun phrases (e.g., nouns, proper name, personal pronouns, etc.). Since predicates (e.g., verbs and verb phrases) cannot name, they must be turned into names by supplying an appropriate noun. Thus, 'pink' in 'Some pigs are pink' must be replaced by 'pink (pigs' to yield 'Some pigs are pink pigs', adhering to the two-name model. The modern logician (I targeted P. Geach), while rejecting the two-name theory, shares with traditionalists like Carroll, the mistaken view that the only kinds of terms that can play the semantic role of naming are noun phrases. Yet no one, modern or traditional, has ever provided a good reason for holding such a view.

9 It Was His Own Invention: Carroll on Cancellation

Jabberwocky, (18), 1989, pp. 270-31

'Are you ready? This is the driest thing I know. Silence all round, if you please!'

Since 1974 I have been writing about Lewis Carroll's logic in the pages of this journal. Mostly this has consisted of brief notes citing this or that interesting (to me) logical notion of Carroll's, followed by analysis showing that, while he was an imaginative logician, Carroll was not a first-rate one. He often failed to appreciate the full import of his ideas for a system of formal logic. Now I want to show that at least one of his ideas in logic was imaginative, foresighted, simple, beautiful, and well-exploited by him. Perhaps this will make up for the mildly negative things I've said of Carroll the logician over the past several years – for I really do admire his logical work, and, like all logicians, find it (as well as *Alice*) a delightful source of examples.

Logicians worry about arguments – sets of statements in which a claim is made (the conclusion) and evidence is cited for it (the premises). From the time of Aristotle, the father of formal logic, until the time of Carroll the kinds of statements seen as making up an argument were always analyzed as *categorical*. A categorical statement always consists of two parts, a *subject* and a *predicate*. A subject itself consists of two parts, a *quantity* expression (for example, 'all', 'every', 'some') and a term. A predicate consists of a *quality* expression (for example, 'is', 'are', or 'isn't') and a term. Terms are nouns, verbs, adjectives, noun phrases, verb phrases, or adjectival phrases. Since there are two kinds of quantity and two kinds of quality, there are four kinds of categorical statements. Let 'A' and 'B' be any two terms, then the four categorical forms (along with the standard letter-names by which they have been know since medieval times) are:

A: Every A is B

E: Every A isn't B (= No A is B)

I: Some A is B

O: Some A isn't B

Suffice it to say, there is *much* more to be said about such forms, but his should do for now.

Traditional logicians (i.e., those who lived before our century) took all arguments to consist of categorical statements. More specifically, they were generally interested in syllogistic arguments (*syllogisms*). A syllogism is an argument in which the number of statements (premises plus conclusion) equals the number of different terms used in the argument. And, even more specifically, they were interested in the kinds of syllogisms which had been first scrutinized by Aristotle himself – syllogisms consisting of three statements (two premises plus a conclusion) and, thus, containing three kinds of terms. Such syllogisms are *classical*. An obvious and famous example of a classical syllogism (along with the name it has enjoyed since medieval times) is:

BARBARA Every A is B

Every B is C

Hence, every A is C

As you can imagine, since each of the statements of a syllogism can be one of the four categorical forms, there are a large number of possible classical syllogisms. But what logicians want to do (among other things) is pick out from these just the ones that are *valid*. To say of any argument that it is valid is to say that it is logically impossible tor its premises all to be true while its conclusion is false. Suppose I ague, according to the BARBARA pattern, as follows:

Every philosopher is a logician

Every logician is wise

Hence, every philosopher is wise

Were I to assert both my premises (as true) and yet deny my conclusion (i.e., assert it as false, i.e., assert is negation as true), I would contradict myself. I would be asserting, in effect:

> Every philosopher is a logician and every logician is wise, but not every philosopher is wise

All systems of logic take this to be a contradiction, so that the original argument is valid. We always want to argue validly rather than invalidly. This is so because we always want to avoid contradicting ourselves. And, as Tweedledee said, 'That's logic'.

Now there are many ways to do logic. One of the ways in which one logician may proceed differently from another concerns just how the validity or invalidity of a syllogistic argument is to be decided. As long as the argument is relatively simple, as many classical syllogisms are, our unaided common sense is a fairly good guide. But if the argument gets a bit longer (has, say, nine or ten terms/statements) or a bit more complex (contains compound or other complex terms), then we need some easy to use, universally applicable, mechanical procedure which we can apply to the argument to decide about its validity. Such a procedure is called, naturally, a *decision procedure*. And logicians over the centuries have devised a large variety of these. And this is where we come, at last, to Lewis Carroll. He devised a decision procedure for syllogisms of any number of terms/statements. Moreover, the procedure which he used is simple and elegant, and effective. Now I must not exaggerate. I doubt that our hero was fully cognisant of just how important this procedure was. And it is true that the procedure can be extended in many ways of which he was clearly unaware. Nonetheless, Carroll must be given the credit he deserves. It was his own invention.

Of course, as with so many good inventions, Carroll's decision procedure for syllogisms was later reinvented by someone else. In our day the American philosopher and logician, Fred Sommers, has developed a complete and universal syllogistic logic which fully exploits the procedure to a degree hardly imaginable to Carroll himself.[1]

Carroll's procedure is simple and easy to use (a necessary requirement for any logical decision procedure). He introduced his 'Method of Underscoring' in *Symbolic Logic* (Bk. VII, ch. ii, § 3). According to the notation which he had introduced earlier in *Symbolic Logic*, each of the four standard forms of categorical statements can be symbolized as follows:

A:	Every x is y	=	$x_1y'_0$
E:	No x is y	=	xy_0
I:	Some x is y	=	xy_1
O:	Some x isn't y	=	xy'_1

In this notation a prime stroke (') indicates the negation of the expression to which it applies, the subscript '1' indicates that something satisfies any term to its left, and the subscript '0' indicates that nothing satisfies any term to its left. Juxtaposition of terms indicates conjunction. Thus one might read Carroll's formulae as:

A: $x_1y'_0$ = Something is x and nothing is both x but not y = Every x is y[1]

E: xy_0 = Nothing is both x and y = No x is y

I: xy_1 = Something is both x and y = Some x is y

O: xy^1_1 = Something is both x and not y = Some x isn't y

Consider now a standard classical syllogism in BARBARA form.

Every philosopher is a scholar

Every scholar is wise

Hence, every philosopher is wise

Using Carroll's notation it is formulated as:

$p_1s^1_0 \dagger s_1w^1_0$ ¥ $p_1w^1_0$

(The dagger (†) represents the conjunction of the premises, the ¥ sign introduces the conclusion.) Notice that of the three terms found in the premises (p, s, and w) one is absent from the conclusion (s). Notice as well that the eliminated term was used in one premise negated and in the other unnegated. In presenting his method Carroll wrote,

> Let us agree to mark the eliminated letters by *underscoring* them, putting a *single* score under the *first* and a *double* one under the *second*.[3]

Our example above would become, then,

$p_1s^1_0 \dagger \underline{\underline{s}}_1w^1_0$ ¥ $p_1w^1_0$

Such a method can be used to decide the validity of a syllogism (once the underscored terms are cancelled the result will be equivalent to the conclusion if and only if the argument is valid). It can also be used to determine which conclusion validly follows from a given set of premises. Here is one of Carroll's examples.[4] From the following premises draw the logically correct conclusion.

> (1) I trust every animal that belongs to me; (2) Dogs gnaw bones; (3) I admit no animals into my study, unless they will beg when told to do so; (4) All the animals in the yard are mine' (5) I admit every animal, that I trust, into my study; (6) The only animals, that are really willing to beg when told to do so, are dogs.

Symbolising the premises (details of which we spare the reader) yields the following:

(1) (2) (3) (4) (5) (6)

$$\underline{h}_1\underline{b}^1_0 \dagger \underline{c}_1 d^1_0 \dagger \underline{k}^1\underline{a}_0 \dagger e_1\underline{\underline{h}}^1_0 \dagger \underline{\underline{b}}_1\underline{\underline{a}}^1_0 \dagger \underline{\underline{k}}_1\underline{\underline{c}}^1_0 \quad ¥ \quad ?$$

Cancellation of appropriate pairs of underscored terms results in the conclusion: $e_1 d'_0$.

Carroll's method is a powerful one. It can be used to decide the validity of any syllogism, and it can be used to generate logically correct conclusions from any set of categorical premises. Moreover, it is simpler than it looks. A little though shows that the cancelling of negated/unnegated pairs of terms and the subsequent conjunction of remaining, uncancelled terms amounts to nothing more than simple algebraic addition. Carroll's argument above could be easily, and more perspicuously, written as:

$$(-(+h-b))+(-(+c-d))+(-(-k+a))+(-(+e-h))+(-(+b-a))+(-(+k-c)) = -(+e-d)$$

Here negated terms are marked my minus, unnegated terms by plus, Carroll's zero is taken as a minus on the entire statement, daggers are replaced by pluses, and the conclusion sign is replaced by the equal sign. Algebraic simplification (viz., driving external minuses inward to the right) give us:

$$(-h+b)+(-c+d)+(+k-a)+(-e+h)+(-b+a)+(-k+c)$$

i.e., $-h+b-c+d+k-a-e+h-b+a-k+c$

i.e., $-e+d$, which equals $-(+e-d)$, or, for Carroll: $e_1 d'_0$

'I'm afraid I can't put it more clearly,' Alice replied very politely ...

Notes

1 Sommers and his followers have developed this system in many places over the past two decades. See especially: F. Sommers, *The Logic of Natural Language* (Oxford 1982); and G. Englebretsen, (ed.), *The New Syllogistic* (New York 1987).

2 Carroll always took universal affirmations to have tacit existential import. This is revealed in his notation by the use of subscript, 1, on the subject terms of such statements.

3 *Symbolic Logic* (Dover, New York 1985), p. 91.

4 Ibid., p. 123.

Note 2021

In this article, I tried to make up for a few mildly negative things I've written in the past about Carroll's logic. I wanted to show that his theory of logical cancellation was insightful and important. After reviewing the basic elements of traditional syllogistic logic, I focused on the issue of devising a *decision procedure*, a way of determining whether a given argument is valid or invalid (only if it is can it then it can be shown to be valid by a *proof procedure*). Carroll devised a decision procedure that can apply to any argument (no matter how long or complex). Improvements were made by others a century later, but the basic procedure was his. He revealed it in *Symbolic Logic* and called it the "Method of Underscoring". Aware that in any valid syllogism there is a least one term that occurs in each premise but not in the conclusion. Such terms are *cancelled.* By underscoring such terms the method makes it easy and obvious to see any required cancellation. Once it is applied, the decision about validity is made: all properly underscored pairs of terms are cancelled – valid; not all such pairs cancelled – invalid. "It was his own invention."

10 The Mad Hatter Writes a Book

Jabberwocky, (18), 1989, pp. 43-45

Gilles Deleuze. *The Logic of Sense* (Columbia University Press, 1990).

> 'What's the French for fiddle-de-dee?'
>
> 'Fiddle-de-dee's not English,' Alice replied gravely.
>
> 'Who ever said it was?' said the Red Queen.

In 1969 Gilles Deleuze published *Logique du Sens* (Les Editions de Minuit, Paris). Now we have, after twenty years, an English translation. It seems to me that with this new translation the time is ripe for an assessment by those English readers interested in Lewis Carroll – for Deleuze has claimed to derive great inspiration from Carroll, and tries to say a great deal about him in this Book. Deleuze was a philosopher at the University of Paris for many years until he retired in 1987. During those years he established a European reputation based primarily on his work on Nietzsche. As both a philosopher and a Carroll enthusiast myself, I am doubly interested in *The Logic of Sense.*

I suppose in all fairness I should show my cards immediately. So: I believe this is a very bad book – both as a piece of philosophical scholarship and as a source of any insights into Carroll or his work. In particular, I could never recommend it to any of the readers of this journal as anything more than curiosity. Few English readers who love *Alice* will either love or even recognise her as she appears in Deleuze's pages. I first read portions of the French text in the mid-seventies. At that time I had suspicions about Deleuze's strange ideas, but I did not trust my own French enough. Reading the English now confirms all my initial suspicions – and more.

The book is a strange amalgam of literary criticism, psychoanalysis, and philosophy. Even granting that, in contrast to Anglo-American scholarship, the Continental style is never to make any effort at keeping these subjects separate, the result is a singularly complex and confusing book. The general thesis of *The Logic of Sense* is (roughly) this: Some philosophers have looked to the 'heights' (abstractions, ideas, forms, etc.) for philosophical understanding. They tend to follow Plato. Others have looked to the 'depths' (the material world, causes, bodies). They tend to follow the pre-Socratics. The proper locale however, according to Deleuze, is the 'surface' (effects, events, states of affairs), discovered by the Stoics, understood by Nietzsche, exploited by Carroll, where sense meets nonsense. In reading this book we learn such things that 'nonsense enacts a *donation of sense* (69, '...the mistake made by logicians, when they speak of nonsense, is that they offer laboriously constructed, emaciated examples fitting the needs of their demonstrations, as if they had never heard a little girl sing, a great poet recite, or a schizophrenic speak' (83), '...sense is presented both as that which happens to bodies and that which insists in propositions' (125), 'Humor is the art of the surface and of the doubles, of nomad singularities and of the always displaced aleatory point; it is the art of the static genesis, the savoir-faire of the pure event, and the "fourth person singular" – with every signification, denotation and manifestation suspended, all heights and depths abolished.' (141). *The Logic of Sense* is nearly four-hundred pages packed full of this kind of philosophical nonsense!

So what does Deleuze have to say about Carroll? More nonsense for the most part. He does not so much give an exposition or an explication of Carroll's work as a manipulation of it to fit his own bizarre thesis. He does of course pay homage to Carroll. But most of what he has to say about Carroll and his work would not be recognized, accepted, or even understood by most of us. For example, Deleuze claims (9) that Carroll abandoned his original title, *Alice's Adventures Underground*, when he realized that Alice had discovered the 'surface', after rejecting the false 'depths' of the first half of the book (rabbit holes, tunnels, shafts, burrows, etc.). On page 10 we learn that 'Lewis Carroll detested boys in general' and that 'As a general rule, only little girls understand Stoicism'! Those who hope light will be shed on the serious logical problem raised by Carroll in 'What the Tortoise Said to Achilles' will be disappointed by Deleuze's discussion (16), which completely misses the point.

The nonsense about Carroll continues throughout the book. Thus Deleuze tells us that the mathematical and logical works were always signed 'Dodgson'! (22) In a discussion of logical paradox we are told 'What is good for Carroll is good for Logic' (74). One might just as well say 'What is good for logic is good for Carroll'. After warning us about the grave dangers of giving a psychoanalytic critique of Carroll (92), Deleuze plunges into a seemingly endless (186-233) recitation of psycho-sexual perversions. We learn (to our surprise, I think) of such 'Carrollian' themes as orality, anality, schizophrenia, madness, the Oedipus complex, castration, desexualisation, neuroses, sexual impotence, and on and on. Then: 'The squares of the chessboard which must be crossed clearly represent erogenous zones, and becoming-queen refers to the phallus as the agency of coordination.' (236) And: 'Thus Carroll, perverse but without crime, perverse but nonsubversive, stuttering and left-handed, uses the desexualized energy of the photographic apparatus as a frightfully speculative eye, in order to invest the sexual object par excellence, namely the little girl-phallus.' (243)

The few interesting or good *passages* concerning Carroll (e.g., a nice analysis of the Gardener's song in *Sylvie and Bruno* (26), and another of the wonderful exchange between Alice and the Knight concerning the name of the song 'Haddock's Eyes' (29), are too rare to warrant the whole dreary book.

In summary, I could not recommend *The Logic of Sense* to anyone – not to philosophers, not to literary critics, and (most definitely) not to those who care about reading and enjoying (and trying to understand) Lewis Carroll, a man who knew how to take care of the sense while letting the sounds take care of themselves.

> 'If any one of them can explain it' said Alice ... 'I'll give him a sixpence. I don't believe there's an atom of meaning in it.'

Note 2021

This was my review of the English translation of Gilles Deleuze's book *Logique du Sens*. I will say it again now: this is a very bad book. Deleuze enlists Carroll in a larger attempt to support a wild, confused, confusing, unfocused, maddening, *senseless* thesis about .... . Who knows? Suffice it to say that few of us would recognize his version of either Carroll or Carroll's creations. It pained me to write even this brief note, for it forced me to engage with *Logique du Sens* one more time.

11 The Properly Victorian Tortoise

Jabberwocky, (23), 1993/94, pp. 12-13

"Whatever Logic is good enough to tell me is worth writing down."

Two decades ago in the pages of *Jabberwocky* I tried to address the problem raised by Lewis Carroll's "What the Tortoise Said to Achilles' (*Mind*, Vol. IV New Series No. 14 April 1895). I say 'the problem raised', but the fact is that for the past century logicians have not been able to agree on just *what* that problem might be. I was only one of dozens who have tried to get clear about just what Carroll was about with his Tortoise and the hapless Achilles. The main logical outlines of the story are clear enough. The Tortoise aims to show that deduction is impossible once the rules of inference are assimilated to argument premisses. Suppose *A*, *B*, *therefore Z* is a formally valid deductive argument. In order to draw conclusions *Z* from premisses *A* and *B* a further assumption seems required, viz., *If A and B are true, then so is Z*. Call this *C*. So now in order to validly derive *Z* one must grant *C*, in addition to *A* and *B*. But this new, augmented argument (*A, B, C, therefore Z*) will itself require a further hypothetical assumption (say *D*), etc., etc, *ad infinitum.*

Carroll's puzzle shows that one ought not assimilate rules to premisses. The trouble is that no logician ever did assimilate rules to premisses in the way the Tortoise did. Thus, the long history of logicians trying to figure out just why Carroll formulated the puzzle. Why poke fun at Tortoise Logic if no logician ever espoused it? Back in 1974 I argued that Carroll's real target was not the Tortoise Logic, which assimilates rules to premisses, but the 'Turtle' Logic, which, conversely, assimilates premisses to rules. Both theories put into a single cage beasts which must be carefully kept apart. He poked fun at the Turtle indirectly by poking fun as his cousin the Tortoise. He knew no logician ever followed the latter, but there *were* logicians who followed the former – J.S. Mill, in particular.

Now, though I have certainly grown older and, presumably, wiser, I have not changed my mind about that interpretation of Carroll's story. But I have continued to think about the Tortoise and what he said to Achilles, and I believe there is more to say about it. I outlined the Tortoise's position above in the same general terms as used by most of those who have commented on it. In particular, we all begin by saying something like: 'Suppose *A*, *B*, *therefore* *Z* is a valid deductive argument ...'. But Carroll himself (and most of his contemporaries) would *not* have described things quite this way. Twentieth century logicians think of logic as, primarily, a tool for determining which arguments are valid and which are not, and for proving those which are. So they naturally read Carroll as being concerned with arguments for which the premisses and conclusion are given. However, Nineteenth century logicians, following Boole, and including Carroll, tended to see logic as a tool for drawing conclusions from stated premisses. In the typical case, for them, premisses where given and logic's task was to discover the strongest conclusion derivable from those premisses. The best evidence for this difference is found by comparing the kinds of problems and exercises found in logic texts published before about 1910 with those found in texts published after that time. *Symbolic Logic* is itself a treasure trove of Nineteenth century logical problems and exercises. So a proper reading of the Tortoise's argument must see it in Nineteenth century terms. The Tortoise is *not* offering a set of premisses and a conclusion and then claiming that the validity of such an argument must depend on the addition of further premisses, *ad infinitum*. Any logician knows that *if* the argument is valid, adding any number of premisses will not remove its validity But this talk of validity is misplaced. It is logicians of our century who emphasize decisions of validity. Logicians of Carroll's century emphasized the drawing of appropriately rich conclusions from given premisses. Read in these terms, the Tortoise was claiming that the set of premisses, *A*, *B* was not sufficient to yield the conclusion *Z* without the addition of an infinite number of other premisses. This is a clearer and historically more accurate way of reading Carroll, and it is still compatible with my old thesis about the Turtle.

Note 2021

This was my second pass at Carroll's "What the Tortoise Said to Achilles." This time I again introduced the idea of Turtle logic. But I went on to highlight a difference between the way Nineteenth Century logicians conceived the task of logic and the way later logicians conceived of it. Carroll and his contemporaries favoured the notion that logic is a tool for drawing from given premisses a conclusion that is revealed by proof. Modern logicians see logic as a tool for determining (and then proving) the validity of an argument in which a given conclusion can be derived from a set of given premisses. Recognizing this difference sheds a bit more light on Turtle logic.

12 Two Important Logical Insights by Lewis Carroll

Reflections on Lewis Carroll: A Collection of New Essays by Various Hands, Fernando J. Soto and Dayna McCausland, eds., Shelburne, Ontario: The Battered Silicon Dispatch Box – Lewis Carroll Society of Canada, 2000, pp. 64-67

"No, it'll never do to ask: perhaps I shall see it written up somewhere."

It is a curious fact that logicians today are particularly fond of using examples and stories from the work of Lewis Carroll, but they almost always draw these from his nonlogical writings (especially the Alice books and *The Hunting of the Snark).* I hasten to add that his logical writings (*Symbolic Logic* and *The Game of Logic)* are not completely ignored by logicians. One does now and then come across some of Carroll's odd and funny arguments from the logical works. But, as a rule, today's logicians find little use for his logic. Of course, Carrollians can hardly complain if a group of scholars find themselves enamoured of the wit and humour to be mined from the rich veins of Carroll's fiction. Nor can they voice genuine disapproval for the fact that some of these scholars have lifted many of his syllogisms from the logical works. The pity is that modern logician, while recognizing and admiring the humour and fun in Carroll, even in his logical writings, have chosen to eschew his logic.

It's a pity, but it's not surprising. Today's logic is the result of a revolution in logical studies that took place during the last few years of Carroll's life and the first few years of this century. The logician Lewis Carroll was not part of that revolution. His logic, traditional logic, was, for all practical purposes, completely replaced by the end of the revolution by what is now known as Classical Mathematical Logic. Now, as it happens, traditional logic, which had been almost in hiding from the time of Leibniz in the seventeenth century to the middle of the nineteenth century had been enjoying a kind of renaissance in the United Kingdom. Boole, De Morgan, Hamilton, Venn, Jevons, (the American Peirce) and other mathematicians and philosophers were well on their way to developing a system of traditional logic that was far more powerful and effective than any earlier system of traditional logic. The quick success of the new mathematical logic, however, completely swamped their program just as it was nearing completion. Lewis Carroll was not the best of these nineteenth century traditional British logicians, He was one of the very last of them – and he was certainly among the most clever and daring of them.

After a century of nearly unchallenged hegemony, mathematical logic is beginning to face critics who hold that the traditional logic (in spite of its having held the field itself unchallenged for nearly two and a half millennia) has yet to have its best version fully aired. Indeed, there are today a number of logicians who have built the kind of traditional logic envisaged by those nineteenth century traditionalists, and who argue that the resulting system has a number of important advantages over the standard mathematical system. As a part of this new development, there is the beginning of a new appreciation of some of the ideas of Boole, De Morgan, Hamilton, Jevons, Venn, and, especially, Peirce. And, perforce, some of these latter-day traditionalists are rediscovering some valuable ideas in Carroll's version of traditional logic.

Here is one of those ideas. In the third section of the Appendix to *Symbolic Logic* Carroll offered an account of logical negation that was not only different from the one then being touted by the new mathematical logicians but different as well from the view then held by most traditional logicians. Consider a sentence such as "Alice is not afraid." Logicians worry about just what is being negated here by the use of "not" There are at least three possibilities: "afraid" is being negated (thus the paraphrase "Alice is unafraid" – here logicians say that "unafraid" is being affirmed of Alice), "is afraid" is being negated (thus the paraphrase "Alice isn't afraid" – here we say that "afraid" is being denied of Alice), or the entire sentence is being negated (thus the paraphrase "It is not the case that Alice is afraid" – we say the sentence is the contradictory of "Alice is afraid"). Mathematical logicians recognize only the third option. All logical negation is sentential. The typical traditional logician recognized all three. Carroll recognized two kids of logical negation, arguing that that the difference between the first two above is merely a matter of taste. Indeed, he went on to urge that one choose the first type, which makes use of a negated term (e.g., "unafraid"), since logicians of his day tended to shun the use of negated terms wherever they might occur. As he said,

> The fact is, "The Logicians" have somehow acquired a perfectly *morbid* dread of negative Attributes [terms], which makes them shut their eyes, like frightened children, when they come across such terrible

> Propositions as "All not-x are y"; and thus they exclude from their system many very useful forms of Syllogisms.

Many of the logicians I mentioned earlier, who aim to retrieve and strengthen the traditional logic in the face of a century of successes in mathematical logic, hold a view of logical negation not unlike Carroll's.

Here is another of those Carrollian ideas worth recovering from his logic. In "Method of Underscoring" in *Symbolic Logic* (Book VII, chapter ii, section 3) Carroll presented the first effective decision procedure for syllogisms involving any number of terms (where even entire sentences can be construed as terms). The procedure is both powerful and theoretically simple. I offer just one brief example to give the flavour of the technique. Consider a syllogism of the form "Every A is a B. Some C is an A. So, Some C is a B." Carroll used the prime stroke (viz. ') to indicate negation and a subscribed "0", the sign of emptiness. For example, "No S is a nonP" would be written as: SP'_0 indicating that the second term is negated and that the intersection of the two sets S and nonP (indicated by juxtaposing the two terms) is empty, that is, nothing is both S and nonP, which, when symbolized, happens to be logically equivalent to "Every S is P." So Carroll's symbolization of our sample syllogism would look like this: $AB'_0+CA_1 - CB_1$ (I have substituted the plus and equal signs for the ones Carroll actually used). He then noted that the middle term, A, the one not found in the conclusion, appears once un-negated and once negated. He underscored such terms and then applied the rule that all underscored terms in premises would be eliminated yielding the conclusion (for valid syllogisms) or some formula other than the conclusion (for invalid syllogisms).

Carroll was able to show that the procedure was powerful enough to discern validity for any such inference no matter how many terms are involved, and it could be used to determine the missing statement for any enthymeme (inference in which one of the premises or the conclusion is unstated).More importantly, he revealed that one could devise a symbolic algorithm for inference that takes the algebraic sum of the premises to equal the conclusion of a valid syllogism. This very idea is now the heart of the system of revitalized traditional logic now enjoying the greatest degree of success as a alternative to standard mathematical logic.

Carroll once referred to himself as nothing more than an obscure writer on logical matters. The obscurity may be an exaggeration, but the lack of extensive serious attention to his logical work cannot be denied (a similar fate befell most of those late nineteenth century traditionalists I mentioned earlier). I have only given a brief sketch of just two of Carroll's valuable (more than just noteworthy) logical ideas. But there were many others. He had interesting insights into a wide variety of issues in the areas of logic and the philosophy of language, some of which have been recognized to various extents, others of which have not. Thus Carroll contributed still worthwhile insights into such topics as existential import, paradox, diagrams, reference, and the nature of deduction.

"Suppose We Change the Subject"

A final, footnotey kind of remark. Those interested, or just curious, about the contemporary developments in traditional logic, especially the developments that tend to preserve (consciously or otherwise) some of Carroll's logical ideas might wish to consult some of the growing literature in this field. The leader here is the philosopher and logician F. Sommers. His *The Logic of Natural Language* (Oxford University Press, 1982) is especially important. Among his many important papers, one should see "Predication in the Logic of Terms," *Notre Dame Journal of Formal Logic,* 31 (1990), pp. 106-126. Sommers and I have produced a textbook version of the revitalized traditional logic: *An Invitation to Formal Reasoning: The Logic of Terms* (Ashgate Publ. Ltd, 2000). I have offered a history, survey, and theoretical account of this version of traditional logic in *Something to Reckon With* (University of Ottawa Press, 1996), as well as a number of short studies of various ideas found in Carroll's logic. These latter have been published in *Jabberwocky* over the past quarter century.

Note 2021

This is a short piece done for a book of essays published by the Lewis Carroll Society of Canada. I gave brief accounts of Carroll's original ideas about the nature of logical negation and his method of underscoring syllogistic terms to facilitate decisions of validity or invalidity. I also tried to give a more favourable assessment of his place in late 19th Century logic. Finally, I added a bit about how more recent logicians have developed a system of term logic as a more powerful version of traditional logic. It is a system that takes seriously the various ideas first advanced by Carroll. I should add here that the quote about Logicians is from *Symbolic Logic* (Part II, Book X, Chapter III).

13 The Logic Pamphlets of Lewis Carroll

Lewis Carroll Review, (44), 2010, pp. 1-5

Review: Francine F. Abeles, editor

The Logic Pamphlets of Lewis Carroll

The Lewis Carroll Society of North America, 2010

ISBN: 978-0-930326-25-8

> "As regards the introduction of absolutely new symbols ..., I think it should be avoided as much as possible. Generally speaking, it is much better to put to fresh uses the familiar symbols of old acquaintance than have recourse to strangers."

As my first logic teacher told me many decades ago, while citing Tweedledee, it never hurts to begin by quoting something from Lewis Carroll.

This is the fourth volume in the series of Carroll's collected pamphlets published by the Lewis Carroll Society of North America, under the general editorship of Charlie Lovett. Two of the previous volumes (collecting Carroll's writings on mathematics and on political issues) were edited by Fran Ableles. As a mathematician, logician and historian of both these disciplines, she is perfectly suited for her editorial role here. Her introductions to each of the three sections into which she has divided the logic writings amount to as good an overall account of Carroll's contributions to logic as one is likely to find anywhere. As well, Abeles has wisely chosen to include here not only Carroll's own contributions to logic but some of his correspondence and published responses to those contributions as well. And that is important because there is a view held by many logicians now that Carroll's connections to the community of logicians in the late nineteenth century was minimal at best. This collection should dispel that idea once and for all.

Casual readers, as well as serious scholars, can approach Lewis Carroll in a wide variety of ways, concentrating, for example, on his biography, the *Alice* books, photography, games and puzzles, mathematics or logic. He was a many-faceted man. Knowing Carroll only from one side (say, via *Alice*) is like knowing DaVinci only through the Mona Lisa. Still, logic was of immense import to Carroll throughout his adult life. It influenced his fiction as well as his scholarship. Moreover, Carroll came to devote more and more time and effort to the subject as he grew older. This is attested to not only by his books on logic (*The Game of Logic*, 1886/7; *Symbolic Logic*, 1896) but by the ever increasing rate of correspondence, pamphlets and publications during his last two decades.

The systematic study of logic has a very long history, many periods of which were far more interesting and eventful than one might imagine. That history was initiated by Aristotle when he wrote *Prior Analytics*. There Aristotle laid down the general requirements for reasoning in a systematic way, untouched by the rhetorical or emotional. The idea was that reasoning, when properly carried out, will lead from what is taken to be true (the premises) to some new truth (the conclusion). But more importantly, and this is the key, there are certain *forms* of such reasoning that will guarantee this. Aristotle's 'formal logic' is the *syllogistic* and rested squarely on that idea. Syllogistic logic sees reasoning, the inferring of a statement from other statements already taken to be true, to be a matter of the forms of those statements. And the form of a statement was taken to be determined by the arrangement of its *terms*. Virtually all formal logic, from Aristotle to the end of the nineteenth century was, in this sense, a *term logic*. In the middle of that century a number of mathematicians and logicians (most notably Boole) sought to provide a more rigorous treatment of term logic by formulating it in mathematical terms. These so-called algebraic logicians made a number of valuable contributions to logic, but by the turn of the next century they were almost completely eclipsed by a new and radically different way of doing formal logic. This new logic was not a term logic. Rather than taking logical form to depend on the arrangements of terms in statements, the new logic took statements themselves to be basic. This was the most important revolution in the history of formal logic. The victory of the new logic (complete by the early years of the twentieth century) over the old logic was swift and decisive. Today the new logic dominates. Carroll carried out his work in logic in the very midst of that revolution.

The Logic Pamphlets of Lewis Carroll clearly shows how closely involved in the developments of algebraic logic Carroll was. He had a high level of familiarity with the ideas of Boole and his followers (such as De Morgan, Venn, Jevons) and he was involved in a variety of critical exchanges with some of the most important logicians of his day. In many ways, as any Carrollian might expect, his role among his contemporary logicians was as a provocateur. He challenged them with a series of questions, logical puzzles, and paradoxes. Attempts to meet his challenges were numerous and continued after his death (and indeed continue today). Beginning in the early '90s Carroll raised the issue of the proper interpretation of hypotheticals (statements of the form 'if this then that'). He tended to view such statements as involving a relation between the senses of the two terms of the hypothetical, a view contrary to that held by most of his contemporaries (and most logicians today), who hold that the relation involved is a matter of the truth or falsity of the two terms. Carroll produced a number of papers and engaged in significant correspondence concerning this issue. More to the point, he formulated what is usually called the Barbershop Paradox, essentially raising the question of just what the logical constraints are on a pair of hypotheticals whose consequents (the 'then' parts) are mutually contradictory (say, negations of one another). He published the so-called paradox in his paper 'A Logical Paradox', *Mind,* 1894, and it elicited a large number of responses, both published and in correspondence. Abeles has included most of this material here, and a careful study of it shows how Carroll's views of the logic of hypotheticals evolved, eventually coming close to the now standard view of such statements as expressing the relation of material implication (best seen in Russell's response to Carroll's paradox).

Another way in which Carroll challenged his contemporaries was his insistence on interpreting universal statements (those of the form 'Every/no...') as having 'existential import'. The question here is: If I assert that every X is Y, am I committed to the claim that Xs exist? It is unclear that he was consistent in his response. At one point, at least, he wanted to hold that there was no demand for existential import. But, for the most part, he formulated universal statements in such a way that their existential import was explicit, taking, for example, 'Every X is Y' to be elliptical for 'Some X are Y and no X are not Y'. The standard view today is that universals do not have existential import, but it is interesting to note that Carroll's interpretation did not completely disappear. A sophisticated version of it was propounded, for example, in the 1950s by the famous Oxford philosopher P. F. Strawson. He argued that to assert that every X is Y is to assert both that no X are not Y and, if not to assert then at least to 'presuppose', that Xs exist. The year after he published 'A Logical Paradox' in *Mind* Carroll published (also in *Mind*) 'What the *Tortoise Said to Achilles*'. There he challenges logicians to consider the basis for their assumption that the premises of an inference are distinct from the rules of derivation used to reach the appropriate conclusion from those premises. After all, among the things someone making such a derivation must know are included *both* the rules and the premises. So why not, as the Tortoise counsels, take the rules to be additional, but suppressed, premises? Carroll's essay is humorous and clever; and it is also subtle and important for logical theory. More than a century after its publication it still attracts the serious attention of many logicians; a growing body of critical studies surrounds the ideas Carroll adumbrated there. (I only regret that Abeles was unable to include here one or two of the most important of these.)

Perhaps the most important way in which Carroll challenged logicians was by his ability to imagine, invent, formulate, contrive a variety of formal devices and procedures for mechanizing logical practice. Three of these are of particular importance. First, he developed a formal language that functioned as a logical algorithm by virtue of such features as his method of subscripts and primes (indicating quantity and negation respectively) and his method of underscoring (later, barring) that permitted an easy way to note terms for elimination. Latter-day term logicians have formulated similar devices in order to build exceptionally powerful formal systems. Second, he invented a truly beautiful (as well as powerful and useful) system of diagrams for syllogistic inference. Carroll diagrams go beyond Venn diagrams, allowing the representation of any number of terms, and do so in a relatively clear and perspicuous manner. Finally, and perhaps most importantly, Carroll invented his tree method for resolving inferences (especially very long ones, known as sorites). The method amounts to a version of *reductio ad absurdum*, laying out sets of statements on branches of a tree in such a way that descending branches are deduced from higher ones until a contradiction is reached for each branch. The method is extremely powerful and foreshadows the semantic tableau (attributed to E. W. Beth) commonly used today.

This volume is a treasure trove for those logicians who care about the history of their subject (small as their numbers might be). It brings together so much material, both published and unpublished, not easily accessed until now. The items collected in the second section, consisting mostly of Carroll and his colleagues and critics' debates on the Barbershop Paradox are especially valuable. Along with his two published books on logic and W.W. Bartley's edition of *Symbolic Logic, Part II*, this collection provides scholars with most of the essential documents of Carroll's logic *oeuvre*. We are now in a much better position to offer a more accurate assessment of his position in the history of logic.

As that assessment proceeds, how are Carroll's contributions to logic likely to be viewed? Already, compared with other of his contemporaries, Carroll is seen favorably by today's logicians. But this is primarily because of their appreciation of his humour, which is a common feature of the many logical examples he offered. As I see it, there are some telling contrasts between Carroll (along with most of his contemporaries) and today's logicians. Nineteenth century logicians tended to put a much higher value on the development and use of diagrammatic methods in logic and they placed a high value on integrating logical research and pedagogy. But things are changing (don't they always?). Interest in diagrammatic reasoning among mathematicians, computation theorists, cognitive psychologists, and logicians is increasing. Carroll developed an original and powerful system of logic diagrams. The increased interest in how best to teach logic is slower, but it is taking place. As this collection shows, logical pedagogy was a driving force in Carroll's logical researches. Finally, there is now a renewed interest in term logic. New versions, more powerful than those formulated by the algebraists are being built. Carroll developed a number of technical devices (symbolic algorithms and notations) that turn out to be precursors of some of the techniques found in these new systems.

I have emphasized the value Abeles's collection will have for logicians and, especially, historians of logic. But even if you happen not to be one of those, a little logic is always a good thing. As Carroll insisted, logic can be fun (even a game) and is not beyond the reach of anyone who has attained the age of reason.*

* I am gratefully indebted to Amirouche Moktefi for guiding me away from some potential missteps.

Note 2021

I was eager to read and review Fran Abeles's important, expertly conceived collection of Carroll's pamphlets, notes and much of his correspondence concerning logic. Her collection is, and will remain, as I said originally, a "treasure trove" for logicians and historians of logic. In particular, it reveals just how much Carroll's thinking about logic was found in his (until now, insufficiently understood or appreciated) extensive correspondence with many of his important contemporary logicians. This is now an essential contribution to those scholars interested in logic, its history, Carroll's place in that history, and the puzzles he left behind that will continue to challenge logicians for a very long time.

14 The Dodo and the DO: Lewis Carroll and the Dictum de Omni[1]

The Carrollian, (25), 2014, pp. 29-37

The period from the time of Boole in the mid-nineteenth century to the time of Gödel in the 1930s was one of the most exciting and productive in the history of logic. Lewis Carroll carried out his most important logical work in the very middle of that era. In *Symbolic Logic* (1896/1977) he presented a system of formal logic that could be used as a machine for reasoning. Not much originality there. Logicians from Leibniz on have had similar goals. Moreover, while Carroll produced his book in the midst of the greatest revolution in logic since its birth in Aristotle's *Prior Analytics*, his own system seems to have been only lightly touched by the new ideas. Carroll produced not a treatise on logic but a textbook. He was at least as interested in the teaching of the subject as in the subject itself. In this respect his work was more in line with other nineteenth century British logicians (e.g., De Morgan, Jevons, and Venn), who coupled the quest for a broader and more rigorous logic with an eye to pedagogy. This accounted in part for their interest, along with their American counterpart Peirce, in developing systems of logical diagrams. In the twentieth century, after the revolution, logic research and logic teaching grew farther apart. What was original in Carroll's text was his formulations of a number of technical devices (some clever, others less so) for his system (cf. Abeles 2005 and 2007).

Boole took his primary rule of deduction (viz., *equals can be substituted for equals*) to be a version of the traditional *dictum de omni* (cf. Corcoran and Wood 1980, 615-616). Suffice it to say that his rule was not the classic *dictum*. But it does at least preserve the insight that it is a *substitution* principle (though not exclusively of equals for equals). What Boole did was take elimination (of middle terms in a syllogism) to amount to the elimination of a variable from a pair of 3-variable equations. Indeed, he took syllogistic inference to be a matter of equation solution rather than proposition deduction (Corcoran and Wood 1980, 619-624). As is well known now, Boole's notion of elimination is not adequate. In part this was due to his clumsy treatment of particular statements, formulated using his notorious 'v' symbol (cf. Green 1991, 2). It was not until the 1890s, when logicians such as Ladd-Franklin and then Venn recognized two distinct copulae, that particulars were put on a comparable footing with universals.

Consider a Barbara syllogism of the form 'Every A is B; every B is C; therefore, every A is C'. What's the best way to describe how we get from the premises to the conclusion? The two instances of the middle term, 'B', are *eliminated*? The major, 'C', is *substituted* for 'B' in the first (minor) premise? Elimination or substitution? Each is an accurate description of what is going on when we draw conclusions from the premises of such valid syllogisms. But the real question about deductive inference isn't that of *what* is being done but of *how* it is being done. We don't want just a description; we want an account, an explanation. For example, Kepler's three laws describe (closely enough) planetary motion; Newton's theory of gravity explains it. Many people before him described species evolution; Darwin's theory of natural selection explains it. More immediately, most people know how to multiply; not many can explain how it's done.

So what about Barbara? I want to claim that the elimination of middle terms is a shorthand way of describing the results of term-substitution. More importantly, I want to claim that the kind of term substitution that takes place here can best be explained as the result of applying (when appropriately licensed) a rule, derived from Aristotle (*Cat.* 1b9-15, 25b32-35, 32b-33a5, *Pr.* An. 24b26-30), which traditional logicians took as the fundamental principle of syllogistic reasoning. The rule is known as the *dictum de omni et nullo*, usually just called the *dictum de omni*, or even just the *dictum*. I'll refer to is as DO. A standard scholastic formulation was: *Quod de aliquo omni dicitur/negatur, dicitur/negatur etiam de qualibet eius parte.* Or, as Aristotle would say, "What is said of something is likewise said of what that something is said of." Or, as I want to put it, "A term affirmed/denied of a universal subject can be affirmed/denied of any subject that the first subject's term is affirmed of." In effect, DO allows us to substitute the predicate of an A or E categorical for its subject-term in any other proposition in which that subject-term occurs undistributively. It is the universal premise (A/E) that *licenses* this substitution.

Before continuing we need to get clear about the concepts of *distribution* and *license.* When the nineteenth century algebraic logicians began to build and refine their versions of term logic they did not make use of some of the innovations of the medieval logicians. In particular, they made little use of the notion of distribution (Ladd-Franklin was an exception). Unlike the predicate calculus which was introduced by Frege, a term logic flounders without an account of distribution. While there are (and always have been) controversies concerning various details of the doctrine of distribution, its elementary parts are easily stated.

1. A term universally quantified in a proposition is distributed in that proposition.

2. A term particularly quantified in a proposition is undistributed in that proposition.

3. A term affirmed of a subject is undistributed.

4. A term denied of a subject is distributed.

5. A term distributed/undistributed in an expression is undistributed/distributed in the negation of that expression.

There followed distribution rules governing syllogistic validity, such as that the middle term must be distributed at least once and that majors and minors must preserve their distribution values (distributed or undistributed) from premises to conclusion.

As it happens, a powerful, but natural, term logic can be formulated taking all syncategorematic expressions as signs of distribution value, with '-' for distributed and '+' for undistributed (cf. Sommers and Englebretsen 2000; also see Makinson 1969, Williamson 1971; Sommers 1975; Geach 1976; Katz and Martinich 1976; Rearden 1984; Englebretsen 1985; Parsons 2006).

Ryle interpreted "What the Tortoise Said to Achilles" (Carroll 1895) as a lesson about natural laws (which are stated in the form of universal propositions). An infinite regress would be initiated if one were to confound them with premises in scientific deductions. Instead, such laws are, according to Ryle (seemingly following a suggestion by Ramsey), to be taken as "inference licenses," which are to scientific explanations what logical rules are to formal deductions. The project of accounting for scientific explanation has advanced at an accelerated rate since Carroll's day and I won't get involved in it now–besides, I don't think it's what Carroll was worried about anyway. Still I do want to borrow Ryle's notion of *license*, confined now just to the field of logical inference. As I read Carroll (cf. Englebretsen 1974 and 1993/4), one needs to be diligent in distinguishing argument premises from inference rules. Treating the latter as premises ("Tortoise logic") or treating the former as the latter ("Turtle logic") leads to serious logical trouble. I want to urge that a license is likewise to be distinguished. In particular:

1) Every license is a universal premise. [the Englebretsen principle (he said modestly)]

1.1) No particular premise is a license.

2) Not every universal premise is a license.

3) No premise is a rule. [the Carroll principle]

3.1) No license is a rule. [which happens to follow from 1 and 3 by DO]

I've said that DO applies whenever it's licensed. The license for DO is a universal (A/E) premise. An inspection of the four perfect first figure syllogisms reveals that in each case the major premise is a universal and the conclusion is the result of substituting the predicate (quantifier plus predicate-term) for the middle term in the minor premise. Note that in each case the middle term of the minor premise is undistributed and the syllogistic rules of distribution hold. Since all valid syllogisms are reducible to those of the first figure, it follows that all (and only) valid syllogisms license DO.

More often than not the suppressed premise of an enthymeme is the license. Licenses are always universal and are often necessarily true. The latter characteristic invites suppression–such a proposition "goes without saying." Suppose I argue that Tom is not married because he's a Catholic priest. The hidden premise, the thing that doesn't need to be said is 'No priest is married' (in other words, 'Every priest is unmarried'). This is the license that, by DO, permits us to substitute 'unmarried' for 'priest' in the explicit premise (where 'priest' is undistributed) to yield the conclusion.

DO, seen as a rule for term substitution when licensed, applies even where the middle term occurs as part of a more complex sub-sentential expression. Consider the argument: 'Every logician is a philosopher; every admirer of all logicians is a fool; therefore, every admirer of all philosophers is a fool'. The proof of this reveals a double reliance on DO. Let the first premise be line1. Line 2 is the tacit, but innocuous tautology: 'Every admirer of all logicians admires all logicians'. Now line 1 licenses the substitution of 'philosopher' for 'logician' in any proposition making use of an undistributed 'logician'. In the second line 'logician' occurs twice, undistributed the first time and distributed the second. So line 1 licenses the requisite substitution in line 2 to yield, by DO, line 3: 'Every admirer of all philosophers admires all logicians'. Now the second explicit premise ('Every admirer of all logicians is a fool') is line 4. This line is a second license. It permits the substitution of 'fool' for the complex term 'admirer of all logicians' in any proposition making use of an undistributed 'admirers all logicians'. And that term does indeed occur undistributed in line 3 ('Every admirer of all philosophers admires all logicians'). So line 4 licenses the requisite substitution in line 3 to yield, by DO, line 5, the conclusion, 'Every admirer of all philosophers is a fool'.

Carroll, the Dodo, made little mention of DO. He did talk of "eliminands," marking the distribution values of middle terms for their impending elimination. It's not always obvious that middle term eliminands have opposite distribution values, but his method of underscoring amounted to a device for making clear just such a distinction. A good notation makes hidden things obvious. That's why Russell said that a good notation is like a live teacher. For example, De Morgan's "spicular" notation failed to reveal many of his most penetrating logical insights (Merrill 2005). Carroll himself seems to have faced some difficulty because of his notation. In his formal language there is no way to indicate the subject/predicate distinction since it parses propositions as affirmations or negations of term-conjunctions (Bartley attributes Carroll's adherence to existential import to this notational feature: Bartley 1977, 35). But Carroll saw his own notation as at least simpler than Boole's. Let's compare his logical notation with some of his contemporaries. Consider the formulation of Ferison according to Ladd-Franklin, Mitchell, the latter Booleans, and Carroll, respectively:

L-F)		M)		B)		C)	
	$(xy)\bar{v}$		$(\bar{x}+\bar{y})_1$		$(xy)=0$		xy_0
	$(xz)v$		$(xz)_u$		$(xz)\neq 0$		xz_1
	$\therefore (z\bar{y})v$		$\therefore (z\bar{y})_u$		$\therefore (z\bar{y})\neq 0$		$\therefore zy'_1$

Only in M can we discern the distribution values of terms (taking Mitchell's over-scoring as a mark of distribution and absence of over-scoring as a mark of nondistribution). Carroll, of course, would use his method of underscoring to show that the middle was eliminable. But his underscoring was not part of his formulation; it was a note of reminder. However, notice that the notations '=0', '≠0', 'v', '$\bar{v}$', '1' and '0' are all sentential operators. This is obvious for L-F and B, and Carroll makes it clear that a subscript is meant to apply to the entire expression to its left (*Symbolic Logic*, Part I, book VI, chapter 1).

Let's prefix Carroll's subscripts to the entire proposition so as to reveal their sentential scope:

C.1) 0(xy)

1(xz)

∴ 1(zy′)

Since '0' and '1' are inverses, we can reformulate them as '-' and '+' respectively.

C.2) -(xy)

+(xz)

∴ +(zy′)

Carroll's prime sign is a unary term negator (*Symbolic Logic*, I, III, 1; also II, X, 3), so we can replace it with a prefixed minus.

C.3) -(xy)

+(xz)

∴ +(z(-y))

Finally, since juxtaposition is taken as conjunction, it can be denoted by a binary addition sign.

C.4) -(x+y)

+(x+z)

∴ +(z+(-y))

This syllogism could be read as: 'It's not the case that something is both x and y; it is the case that something is both x and z; therefore, it is the case that something is both z and not y'. An inspection of C.4 shows that the plus and minus signs amount to signs of distribution values. Terms in the range of an odd number of minuses are distributed; the rest are undistributed. We could even drive in the external signs (simultaneously suppressing redundant unary pluses) to yield:

C.5) -x-y

+x+z

∴.+z+(-y)

Notice that now no underscoring is required. Simply adding the premises eliminates the middle and produces the conclusion. Better still, we could "reduce" our Ferison to the perfect Ferio by simply converting the minor and applying obversion to the conclusion. Thus:

C.6) -x-y

+z+x

∴ +z-y

In *Symbolic Logic* (II, X, 8) Carroll gives an example of a syllogism that is first reduced to Fresison (by obversion on the minor) and then is reduced to Ferio (by simply converting each premise). He then writes, "The validity of Ferio follows directly from the Axiom '*Dictum de Omni et Nullo*'" (Bartley 1977, 247).

A natural way to look at C.6 is as a pattern for elimination. There in the premises are a couple of xs that get eliminated once the conclusion is drawn. A more illuminating way to see C.6 is as an algebraic addition. The premises add up to the conclusion; i.e., the negative and positive occurrences of x cancel one another. More illumination still is shed by seeing such forms as exhibiting the substitution of one expression for another – DO. In C.6 the major denies y of every x, the minor then affirms x of some z. In the conclusion the x of the minor has been replaced by the minus y of the major. In general, the predicate of a universal proposition can replace the subject-term of that proposition in any other proposition in which that subject-term occurs undistributed. Elimination of minor terms is just a short-hand way of describing the *results* of applying DO, which is the principle that accounts for term-substitution when licensed by an appropriate universal proposition. Notice that this kind of application of DO (licensed substitution) is what we tend to do quite naturally (even if we overdo it and commit fallacies from time to time).

Here's a valid sorites that yields its conclusion by the application of DO three times.

1. Every boy loves a girl.
2. Every girl adores a cat.
3. All cats are mangy.

4. Whoever adores something mangy is a fool.

7. So, every boy loves some fool.

Notice that the expression 'adores a cat' is affirmed of 'every boy' in 2, and 'girl' occurs undistributed in 1, so we can substitute 'adores a cat' for 'girl' in 1 to give us, by DO,

5. Every boy loves someone who adores a cat.

Applying DO (licensed by 3), we can substitute 'mangy' for 'cat' in 5 to yield

6. Every boy loves someone who adores something mangy.

Finally, applying DO once more, we see that 4 licenses the substitution of 'fool' for 'adores something mangy' in 6 to yield the conclusion.

Carroll, like his contemporaries, recognized the importance of DO. And like them, he noted that the correct application of the dictum results in the elimination of middle terms. Unlike most of his contemporaries, he made use of a notational device (underscoring) to mark the opposing distribution values for middle term in preparation for their mutual cancellation. What neither he nor his contemporaries succeeded in doing was to offer an explanation of the dictum. I've tried to indicate what that explanation might look like.

[1] Portions of this essay were presented to a joint conference of the British and Canadian Societies for the History and Philosophy of Mathematics at Concordia University in Montreal on 29 July 2007. I want to thank the participants of the special session on *Lewis Carroll: Logician*, especially Fran Abeles and Amirouche Moktefi as well as the organizers.

Bibliography

Abeles, F. 2005. "Lewis Carroll's Formal Logic," *History and Philosophy of Logic*, 26:33-46.

__________ 2007. "Lewis Carroll's Visual Logic," *History and Philosophy of Logic*, 28:1-17.

Bartley, W.W., ed. 1977. *Lewis Carroll's Symbolic Logic*, New York: Clarkson N. Potter; 2nd edition, 1986.

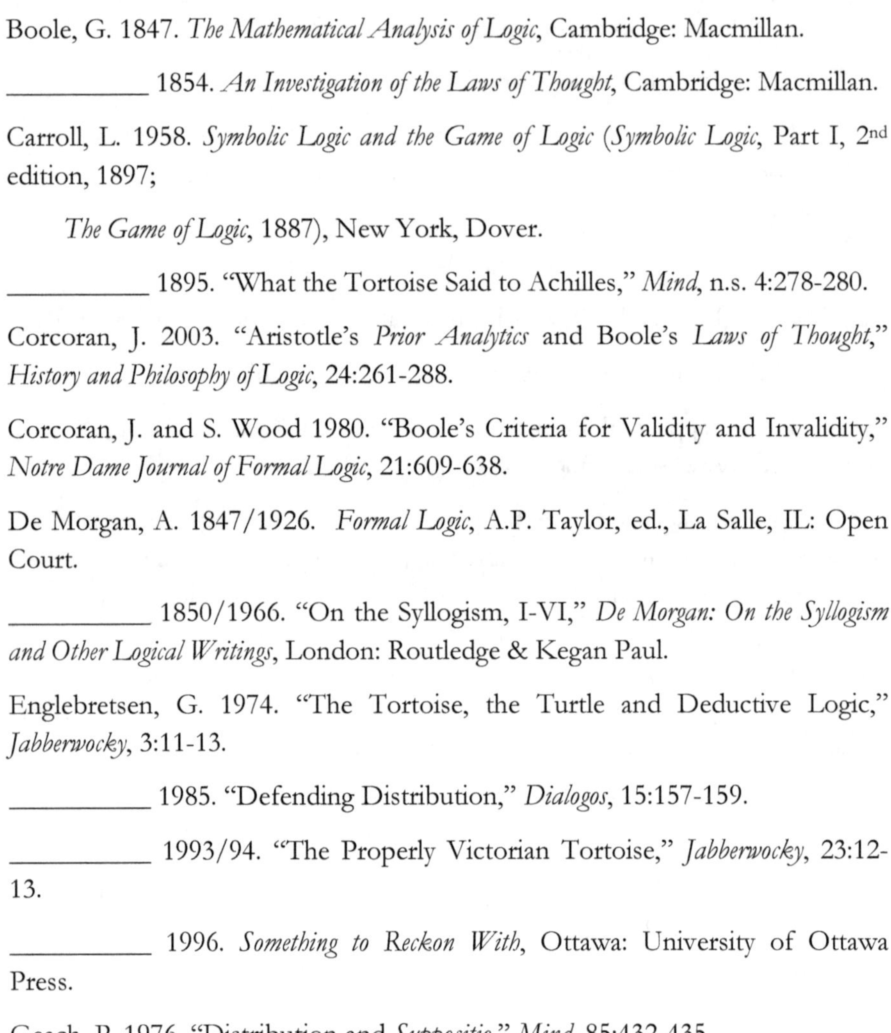

Boole, G. 1847. *The Mathematical Analysis of Logic*, Cambridge: Macmillan.

__________ 1854. *An Investigation of the Laws of Thought*, Cambridge: Macmillan.

Carroll, L. 1958. *Symbolic Logic and the Game of Logic* (*Symbolic Logic*, Part I, 2nd edition, 1897;

The Game of Logic, 1887), New York, Dover.

__________ 1895. "What the Tortoise Said to Achilles," *Mind*, n.s. 4:278-280.

Corcoran, J. 2003. "Aristotle's *Prior Analytics* and Boole's *Laws of Thought*," *History and Philosophy of Logic*, 24:261-288.

Corcoran, J. and S. Wood 1980. "Boole's Criteria for Validity and Invalidity," *Notre Dame Journal of Formal Logic*, 21:609-638.

De Morgan, A. 1847/1926. *Formal Logic*, A.P. Taylor, ed., La Salle, IL: Open Court.

__________ 1850/1966. "On the Syllogism, I-VI," *De Morgan: On the Syllogism and Other Logical Writings*, London: Routledge & Kegan Paul.

Englebretsen, G. 1974. "The Tortoise, the Turtle and Deductive Logic," *Jabberwocky*, 3:11-13.

__________ 1985. "Defending Distribution," *Dialogos*, 15:157-159.

__________ 1993/94. "The Properly Victorian Tortoise," *Jabberwocky*, 23:12-13.

__________ 1996. *Something to Reckon With*, Ottawa: University of Ottawa Press.

Geach, P. 1976. "Distribution and *Suppositio*," *Mind*, 85:432-435.

Green, J. 1991. "The Problem of Elimination in the Algebra of Logic," *Perspectives on the History of Mathematical Logic*, T. Drucker, ed., Boston: Birkhäuser, 1-9.

Katz, B.D. and A.P. Martinich 1976. "The Distribution of Terms," *Notre Dame Journal of Formal Logic*, 17:279-283.

Ladd-Franklin, C. 1883. "On the Algebra of Logic," *Studies in Logic*, by Members of the Johns Hopkins University, Boston: Little, Brown & Co., 17-71.

Makinson, D. 1969. "Remarks on the Concept of Distribution in Traditional Logic," *Noûs*, 3:103-

108.

Merrill, D.D. 2005. "Augustus De Morgan's Boolean Algebra," *History and Philosophy of Logic*, 26:75-94.

Mitchell, O.H. 1883. "On a New Algebra of Logic," *Studies in Logic*, by Members of the Johns Hopkins University, Boston: Little, Brown & Co., 72-106.

Parsons, T. 2006. "The Doctrine of Distribution," *History and Philosophy of Logic*, 27:59-74.

Rearden, M. 1984. "The Distribution of Terms," *Modern Schoolman*, 61:187-195.

Ryle, G. 1950. "If, So and Because," *Philosophical Analysis*, M. Black, ed., Ithaca: Cornell University Press

Sommers, F. 1975. "Distribution Matters," *Mind*, 84:27-46.

__________ 1982. *The Logic of Natural Language*, Oxford: Clarendon Press.

Sommers, F. and G. Englebretsen 2000. *An Invitation to Formal Reasoning*, Aldershot: Ashgate Publ.

Venn, J. 1894. *Symbolic Logic*, 2nd ed., London: Macmillan.

Williamson, C. 1971. "Traditional Logic as a Logic of Distribution Values," *Logique et Analyse*, 14:729-746.

Note 2021

This essay was first presented at the conference *Lewis Carroll, Logician* as part of a meeting of the British and Canadian Societies for the History and Philosophy of Mathematics at Concordia University, in Montreal on 29 July, 2007. It first appeared in print in *Proceedings of the Canadian Society for the History and Philosophy of Mathematics*, 20 (2008), pp. 142-148.

Traditional syllogistic logic rests on a basic principle (though controversy remains). The *dictum de omni et nullo* (or simply the *dictum de omni* or even just the *dictum*) says that what holds of (characterizes, is true of) all of something must also hold of whatever that something holds of. For example, what holds of all mammals (being an animal) holds of what are mammals (humans). So, being an animal holds of all humans. What this means is that in any valid syllogism the "middle" term, the term that is in the subject of one premise and in the predicate of the other premise, but not in the conclusion, can be *cancelled*. In *Symbolic Logic*, Carroll devised a notation to make such cancellation (if legitimate) easy to spot. In this essay, I tried to show how the *dictum* amount to a rule of *substitution*. In doing this I introduce another key idea from traditional logic: the doctrine of term *distribution*. Finally, I enlist the notion of an inferential *license*. Taking the *dictum* as a rule governing term substitution, I show *that* such a principle only applies to a given syllogism it is licenses, which, in turn, depends on the one instance of the middle term being distributed and the other being undistributed. Carroll and his contemporaries (e.g., Boole, De Morgan, Ladd-Franklin) each devised systems of logical notation that reveal the central role of the *dictum* and the doctrine of term distribution, which together facilitate term cancellation. None, however, were able to provide a full explanation (beyond just stating it) of how the *dictum* works, namely., when it is licensed.

15 Deriver's License Required

Exploring Topics in the History and Philosophy of Logic

> Mathematicians are a species of Frenchmen: if you say something to them they translate it into their own language and presto! It is something entirely different.
>
> Goethe

We have seen just how important the notion of distribution is to our formulation of the key rule of syllogistic inference, the *dictum de omni*. Indeed, a term logic like TFL can be formulated taking all formative expressions as signs of distribution values, with ' – ' for distributed and '+' for undistributed. Boole took his primary rule of deduction (viz., *equals can be substituted for equals)* to be a version of the *dictum* (Corcoran and Wood 1980, 615-616). Suffice it to say that his rule was not the classic *dictum*. But it does at least preserve the insight that it is a *substitution* principle (though not exclusively of equals for equals). What Boole did was take the elimination (of middle terms in a syllogism) to amount to the elimination of a variable from a pair of 3-variable equations. In fact, he took syllogistic inference to be a matter of equation solution rather than statement deduction (Corcoran and Wood 1980, 616-624).

Consider a simple syllogism of the form 'Every A is B; every B is C; therefore, every A is C'. What's the best way to describe how one would get from the premises to the conclusion? The two instances of the middle term, 'B', are *eliminated?* Each is an accurate description of what is going on when we draw conclusions from the premises of such valid syllogisms. But the real question about deductive inference isn't that of *what* is being done, but of *how* it is being done. We don't want just a description – we want an account, an explanation. For example, Kepler's three laws describe (closely enough) planetary motion; Newton's theory of gravity explains it (closely enough). Many people before him descripe species evolution; Darwin's theory of natural selection explains it. More immediately, most people know how to multiply and divide; not that many can explain how it is done.

So what about our simple syllogism? I want to claim that the elimination of middle terms is a shorthand way of describing the results of term-substitution. More importantly, I want to claim that the kind of term-substitution that takes place here can best be explained as a result of applying (*when appropriately licensed)* the *dictum*. In effect, the *dictum* allows us to substitute the predicate term of a universal affirmative or negative categorical premise for the subject term in any other proposition in which that subject term occurs undistributively. It is the universal premise that *licenses* such as substitution.

Before continuing, we need to get clear about the concept of *license* and say how it relates to the notion of distribution. When the 19th century algebraic logicians began to build and refine their versions of formal logic, they did not make use of some of the innovations of the medieval logicians. With the exception of Christine Ladd-Franklin (Ladd-Franklin 1883), they made little use of the notion of distribution. Unlike the predicate calculus, a term logic flounders without an adequate account of distribution.

As almost logicians (and even some others who have wandered into *Wonderland* or gone through the Looking Glass) know, Lewis Carroll (aka Charles Dodgson) was a mathematician and logician. Near the end of his life he wrote a short dialogue for *Mind* called "What the Tortoise Said to Achilles" (Carroll 1895). A fairly large literature has grown around it. Briefly, what the Tortoise tries to do is get Achilles to admit that in order to draw the conclusion (essentially of the form) 'This A is C' from premises of the forms 'This A is B' and 'Every B is C' he must first add the premise 'If this A is B and every B is C, then this A is C'. Of course, once Achilles allows this he is faced with a new inference (with three premises) and is challenged by the Tortoise to admit still a fourth premise, and then … *ad infinitum*. Though there has been much debate about just what logical point Carroll was trying to make, at least part of the answer is that he was illustrating the importance of avoiding confusing premises with rules of inference. If every rule used to draw a conclusion from a set of premises is also a premise, then an additional rule must be used to draw the conclusion from this augmented set of premises. This new rule must then be added as a further premise, and so on, and so on – an infinite regress. Gilbert Ryle (Ryle 1950) interpreted Carroll as teaching a lesson concerning laws in natural science (which are stated in the form of universal statements). An infinite regress would be initiated if one were to confound them with premises in scientific deduction. These universal statements are to be seen as "inference licenses," which are to scientific explanations what logical rules are to formal deduction. Sidestepping Ryle's take on scientific laws, I want to borrow this notion of *license*, confined now just to the field of logical (viz., syllogistic) inference. Generally, a license is a permission; it permits one to do something under certain specified conditions. A license had, in effect, the form: 'Under such and such conditions, you can do the following …'. By my reading of Carroll (Englebretsen "What Did Carroll Think the Tortoise Said to Achilles?" *The Carrollian,* 28, 2016, pp. 76-83), one does need to be diligent in distinguishing argument premises from inference rules – but a license is likewise to be distinguished from such rules. So:

1. Every license is a universal premise.
2. No particular premise is a license.
3. Not every universal premise is a license.
4. No premise is a rule.
5. No license is a rule.

I've said that the *dictum* applies whenever it's licensed. The license for it is a universal premise. An inspection of Aristotle's four perfect syllogisms reveals that in each case there is a universal premise and the conclusion is the result of substituting the predicate of that premise for the middle term (the term that appears in both premises but not the conclusion) that occurs undistributed in the other premise. Since all valid syllogisms are reducible to those perfect syllogisms, it follows that all (and only) valid syllogisms license the application of the *dictum.* More often than not the suppressed premise of an enthememe is the license. Licenses are always universal and are often necessarily true. The latter characteristic invites suppression – such a proposition "goes without saying." Suppose I argue that Ralph is not married because he is a Catholic priest. The hidden premise, the thing that doesn't need to be said is 'No Catholic priest is married' (or 'Every Catholic priest is unmarried'). This is the license that by the *dictum,* permits the substitution of 'unmarried' for 'priest' in the explicit premise (where 'priest' is undistributed) to yield the conclusion. In the famous example 'Every horse is an animal, so every head of a horse is the head of an animal', the license is the explicit premise. The other premise is the suppressed 'Every head of a horse is a head of a horse' (something too obvious and trivial to need saying). Here one licensed to substitute 'animal' for 'horse' when 'horse' occurs undistributed. In the missing premise, 'horse' occurs twice, but only the second occurrence is undistributed. Substituting 'animal' for that second occurrence of 'horse' yields the desired conclusion.

As we saw, the *dictum de omni*, seen as rule for term substitution when licensed, applies even where the middle term occurs as part of a more complex sub-sentential expression. Consider one more argument: 'Every logician is a philosopher; every admirer of all logicians is a fool; therefore, every admirer of all philosophers is a fool'. The proof of this reveals a double reliance on the *dictum*. Let the first premise be line 1. Line 2 is the tacit, but innocuous 'Every admirer of all logicians admires all logicians'. Now line1 licenses the substitution of 'philosopher' for 'logician' in any premise making use of an undistributed 'logician'. In the second line 'logician' occurs twice, undistributed the first time and distributed the second. So line 1 licenses the substitution of 'philosopher' for 'logician' in line 2 to yield, by the *dictum*, line 3: 'Every admirer of all philosophers admires all logicians'. Now the second explicit premise ('Every admirer of all logicians is a fool') is line 4. This line is a second license. It permits the substitution of 'fool' for the complex term 'admirer of all logicians' in any statement making use of an undistributed 'admires all logicians'. And that term does indeed occur undistributed in line 3 ('Every admirer of all philosophers admires all logicians'). So line 4 licenses the requisite substitution in line 3 to yield, by the *dictum*, the conclusion 'Every admirer of all philosophers is a fool'.

To summarize: to validly draw a syllogistic conclusion, three things are required. First, a universal premise, which acts as a license for substitution. Second, another premise which includes an undistributed token of the subject term of the license. Third, a rule (the *dictum de omni*) that is licensed by the universal premise to substitute the predicate of the license for that undistributed term as it occurs in the other premise.

Note 2021

This item is from my book *Exploring Topics in the History and Philosophy of Logic*, Berlin and Boston: De Gruyter, 2015 (Chapter 4, part 3, pp.74-77). However, I first presented my idea of logical license in terms of distribution five years earlier in "How to Use a Valid Derivers License," *The Reasoner*, 4 (2010), pp. 54-55. In the interim, that old Tortoise refused to leave me alone. So, I followed him (or he followed me) when I turned again to the Carroll's topic of the nature of deduction and saw how the idea of license and distribution could shed some more light.

16 What Did Carroll Think the Tortoise Said to Achilles?

The Carrollian, (28), 2016, pp. 76-83

Whatever *Logic* is good enough to tell me is worth *writing down.*

Then Logic would take you by the throat, and *force* you to do it.[1]

What I Thought I Knew

It seems that every couple of decades or so I feel compelled to return to Lewis Carroll's famous (and, for philosophers and logicians at least, maddening) essay "What the Tortoise Said to Achilles" [Carroll 1895]. In 1974, I argued that the target of Carroll's essay was John Stuart Mill [Englebretsen 1974]. Briefly, my argument was that the infinite regress generated by the Tortoise constitutes a *reductio ad absurdum* argument against one of his assumptions – viz., that rules of deductive inference must be taken as premises (since acceptance of rules as well as premises is required in order to deduce a conclusion). This is "Tortoise logic" and it's highly doubtful that any logician Carroll was aware of was a Tortoise logician. However, there is (appropriately for Carroll) a mirror version of Tortoise logic, and that is what Carroll had in mind. Just as Tortoise logic mistakenly treats rules as if they are hidden premises, "Turtle logic" treats premises as if they are implicit rules. Treating premises as rules is just as mistaken as treating rules as premises ("Why you might just as well say that 'I see what I eat' is the same as 'I eat what I see'!" according to the Hatter). I held, then, that Mill's attack on syllogistic logic was an example of Turtle logic, and thus the target at which Carroll indirectly aimed. I was young; I was foolish; I could have been wrong.

Not so fast! Twenty years later, I continued to defend the view that Mill had been the intended patsy [Englebretsen 1993]. I simply added that in reading Carroll's essay one needs to keep in mind that logicians of the Victorian

---

[1] These quotations are from [Carroll 1895, p. 280].

era would have taken what the Tortoise said about logic in their terms not ours. We tend to see the assessment of the validity or invalidity of deductive arguments as a central logical task; they tended to see the central task of logic as the discovery of the strongest conclusions derivable from a given set of premises. As all logicians (of any species) know, if a deductive argument is formally valid, then adding premises won't change anything. But the Tortoise was a Victorian logician, trying to draw the richest conclusion from the premises; he was not trying to show that the argument is invalid – he was trying to show that no conclusion could be drawn in a finite number of steps. Of course to do so he had to mistakenly treat rules as premises.

Well, that was then, this is now. With age comes caution (and on rare occasions wisdom). What *did* Carroll think the Tortoise said to Achilles? What lesson was the Tortoise supposed to have taught us? I believe Carroll was hunting more than one snark in his essay.

Nineteenth Century Lessons

The first, and most obvious, lesson was, of course, that treating rules of deductive inference as argument premises necessarily results in the absurdity of an infinite regress, making the drawing of appropriate conclusions (and consequently, proof of validity) impossible. His intention here might well have been (as I said in my youth) to offer an oblique rejection of Mill's Turtle logic. But, to be fair, Mill didn't attack syllogistic deduction by claiming that premises are rules. Rather, he charged that such arguments beg the question in that the conclusions of syllogisms are always hidden in *one* of the premises [Mill 1874, esp. pp. 126-157]. Carroll knew this. In *Symbolic Logic, Part I: Elementary*, he wrote: "This formidable objection is refuted with beautiful clearness and simplicity, by these three Diagrams, which show us that, in each of the three Figures, the Conclusion is really involved in the two Premisses taken together, each contributing its share" [Carroll 1897, p. 164]. OK, so I *was* wrong.

Mill was not alone in attacking syllogistic reasoning. Carroll's contemporary and sometimes correspondent, Francis H. Bradley had mounted a sustained attack on such reasoning [Bradley 1883, esp. pp. 245-269]. Bradley's main complaint was that deduction does not lead from given information (provided by the premises) to *new* information (offered in the conclusion), since he held, with Mill (but for different reasons), that any information provided by the conclusion must already be contained in the premises. Victorian logicians in general thought of inference as a process of drawing new information from other information already given or assumed. This concentration on the informational content diminished over the next century (but see [Corcoran 1998]). Nonetheless, I'm not convinced that even Bradley's attack on deduction was Carroll's target. For Bradley was also well-known for his attack on the notion of *relation*. And just like the Tortoise, Bradley made use of an infinite regress argument.

Bradley raised doubts about the idea that any two things could *really* be related [Bradley 1893]. Suppose that two things, *a* and *b*, are related by a relation, R (perhaps *a* is Abelard and *b* is Heloise, and R is the relation of *loves* that Abelard stands in to Heloise – Abelard loves Heloise). For it to be true that *a* is R to *b* there must be, according to Bradley, some single unified thing, a fact, a state of affairs, etc., that accounts for that truth. This thing can't simply consist of three unrelated things (*a*, *b*, and R). For it to be one unified thing (one fact, etc.) its constituents must be bound together by further relations. R may well relate *a* to *b*, but what relates *a* to R? What relates R to *b*? Suppose what relates *a* to R is R1 while R2 relates R to *b*. The regress has begun: what then relates *a* to R1? What relates R1 to R? What relates R to R2? What relates R2 to *b*? What relates R1 to R2? Each purported answer generates a new relation, *ad infinitum*. If relations are the elements that bind relata into units, then they themselves require additional relations to bind *them*. This is just the sort of thing that would have attracted Carroll's interest (after all, he took his characters from the original master of infinite regress – Zeno). It is certainly not unreasonable to suppose that one of the things Carroll had in mind in his essay was to offer an attack on deduction parallel to Bradley's attack on relations in order to show how silly such regress arguments can be. Treating a rule as a premise (thus requiring a further rule, which in turn will be treated as a premise) is like treating a relation between any two things as just another thing (thus requiring a further relation, which in turn will be treated as a thing). It's as silly as thinking that there must be some other adhesive that sticks the glue to the stamp (and even another that sticks the glue to the envelope)!

Other philosophers and logicians have drawn further lessons from the Tortoise. The then-editor of *Mind*, George F. Stout wrote to Carroll claiming that if there is no distinction between affirming a proposition and affirming the truth of that proposition, then "there would appear to be no point in your puzzle," to which Carroll immediately replied that the aim of his "paradox" was not to draw the distinction to which Stout alludes [Bartley 1977, pp. 471-474]. Carroll correctly pointed out that his concern was with distinguishing the antecedent of a hypothetical, the consequent of that hypothetical, and the hypothetical itself as three different propositions. The regress is first generated by recognizing that accepting the consequent depends upon accepting both the antecedent and the hypothetical. Bertrand Russell drew the same distinctions that Carroll had drawn [Russell 1903, p. 35]. But he went on to distinguish *implication*, the logical relation holding between the two unasserted propositions constituting a hypothetical, from *inference*, the logical relation holding between the premises and the conclusion of a valid argument. *That* distinction is not found in Carroll.

Twentieth Century Lessons

Famously, Gilbert Ryle, followed by Stephen Toulmin, drew a lesson about the nature of scientific explanation from what the Tortoise said [Ryle 1950, Toulmin 1953]. Just as rules of inference cannot play the role of premises in arguments in which those rules are used, Ryle argued that natural laws cannot play the role of premises in scientific explanations. Natural laws are not *rules* of inference, but they are "inference licenses" that permit scientific inference to take place at all. Something like this is surely right, and the notion of an inference license is a valuable insight, but it's doubtful that Carroll thought the Tortoise's lesson was about the nature of scientific explanation – rather than about deductive inference more generally. William W. Bartley III, the scholar whose singular contribution to Carroll studies was the discovery, and eventual publication, of the long-lost Part II of Carroll's *Symbolic Logic*, drew a different lesson. Claiming that the interpretation offered by Ryle "is plainly wrong," [Bartley 1977, p. 467], Bartley said that the most plausible interpretation of Carroll's essay was as an attempt by Carroll "to express some difficulties he felt he could not adequately explain" [Bartley 1977, p. 468]. According to Bartley, this inadequacy was due to the faulty logical theory available to Carroll. Presumably the faults here were eventually rectified by developments in logic over the next half-century. Maybe.

Perhaps the most influential logician of the past several decades has been Willard V. Quine. He often made references to Carroll, and, speaking for contemporary logicians in general, admitted that "Some of us do not outgrow him. There are playful absurdities in his tales that tickle the logical mind" [Quine 1981, p. 134]. Quine, a champion of the new mathematical logic had, like Bartley, a fairly low opinion of the kind of old logic represented in the works of Victorians such as Carroll. Nevertheless, he took "What the Tortoise Said to Achilles" seriously. One of the things Quine was anxious to eliminate from the thinking of logicians (especially Rudolph Carnap) was the notion that logic (namely, the set of principles that govern deductive reasoning) is grounded on stipulated conventions (principles generally agreed upon by users) without there being an even more basic logic governing those conventions (Quine favoured a more empirical foundation for logic). It is obvious that if logic *were* grounded on conventions that were themselves grounded on a deeper logic, then an infinite regress is generated. Quine used just such a regress argument in his attack on Carnap, citing Carroll's essays as his model (see esp. "Truth by convention" and "Carnap on logical truth," both reprinted in [Quine 1976]). He learned something from the Tortoise.

James F. Thomson argued that the idea that taking a rule of inference as a hidden premise in that inference leads to an infinite regress "is wrong, and is not established by the story" [Thomson 1960, p. 105]. Thomson rightly points out that adding *any* premise to a valid argument changes nothing – the argument remains valid. But he denies that doing so necessarily leads to an infinite regress. And he's right here as well. So he writes, "The extreme eccentricity of the behaviour of both of the characters may well make us wonder whether Lewis Carroll knew what he was up to in writing the story" [Thomson 1960, p. 99]. Carroll probably had a pretty good idea what he was up to. Adding a premise to an already valid argument never makes that argument invalid (though it always renders an invalid inference valid). In particular, adding a premise that is simply the hypothetical formed by the conjunction of the premises as antecedent and the conclusion as consequent cannot render the argument invalid. The Tortoise wouldn't deny this; Carroll wouldn't deny this; no logician would deny this. However, adding such a hypothetical premise *does* lead to an infinite regress whenever that added premise happens to be an instantiation of the very rule of inference required to draw the conclusion from the other premises. That is why what the Tortoise said to Achilles generated an infinite regress. As Zeno, Carroll and Quine knew, infinite regresses can be vicious.

What I Learned

So, what now, in these latter days, has the Tortoise taught me? One thing is that Carroll's essay was far more subtle and multifaceted than any reader might first (or second, or third) imagine. Some things to learn from the Tortoise: Carroll was a *Victorian* logician, so paying attention to what conclusions can be drawn from a given set of premises was just as important as determining argument validity. Mill probably wasn't a Turtle logician after all; but it's still the case that rules are not premises and premises are not rules. Conclusions are not invariably hidden in premises. Infinite regress arguments are tricky and must be handled with great care. As I've tried to illustrate, others have learned such lessons as well. But, four decades of reading and thinking about Carroll's work on logic have led me to a further lesson that might only have been dimly hidden in what the Tortoise said to Achilles. I have written about this in other places (esp. [Englebretsen 2008]). I'll just briefly state it here without supporting argument. It *is* possible to construe all valid deductive arguments as syllogisms. Every valid syllogism has at least one universal premise (a statement of the form 'Every A is B'). The fundamental principle of syllogistic reasoning is the *dictum de omni*. Over the centuries, the *dictum de omni* has had many friends and probably just as many foes. Among the former, I count Aristotle, Gottfried W. Leibniz, Carroll, myself and many others. The foe with whom Carroll was most familiar was Bradley, who, in his *The Principles of Logic*, referred to the dictum as "vicious" [Bradley 1883, p. 248]. But his credentials as a logician have always been suspect. The dictum allows one to substitute the predicate-term (B) for the subject-term (A) of a universal premise in the other premise whenever the term, A, occurs undistributed in that other premise (on *distributed* and *undistributed* terms, see the next section). The result of such a substitution will be the conclusion. Such a substitution is not allowed (and, thus, validity is not guaranteed) unless there is that universal premise. So, what is required to draw a syllogistic conclusion validly? Answer: three things, (1) a universal premise (following Ryle, I call such a premise a license [Englebretsen 2010]), (2) a second premise in which the subject-term of the license is undistributed, (3) a rule (the *dictum de omni*) for deriving the conclusion by substituting the predicate-term of the license for that undistributed term in the other premise. And, of course, always keep in mind that no rule is a premise and no premise (not even a license) is a rule.

What a clever and provocative teacher that Tortoise was!

My Distribution Lesson

Every term used in any statement is either *distributed* or *undistributed* in that statement. To find out whether a term is or is not distributed in a statement, first paraphrase away all universal quantifiers (expressions such as 'all', 'every', 'each', etc.), leaving only expressions for negation (such as 'not', 'un', 'it's not the case that', 'isn't', etc.), particular quantifiers (like 'some', 'a(n)', 'at least one', etc.), and the terms themselves. Here are some examples:

Every S is P	≡	It's not the case that some S is not P
No S is P	≡	It's not the case that some S is P
Every A is R to some B	≡	It's not the case that some A isn't R to some B
Some A is R to every B	≡	Some A isn't not R to some B

Next, any term in the range of an odd number of negations is distributed, otherwise it is undistributed. Undistributed terms above are P is the first example, R and B in the third, and A and R in the fourth.

Note

I want to thank Francine Abeles and Amirouche Moktefi for a number of extremely insightful comments and recommendations leading, in many cases, to valuable improvements to the original version of this essay

Bibliography

Bartley III, W. W. (ed.), 1977, *Lewis Carroll's Symbolic Logic*, New York: Clarkson N. Potter.

Bradley, F. H., 1883, *The Principles of Logic*, London: Oxford University Press.

--- 1893, *Appearance and Reality*, New York: Macmillan.

Carroll, Lewis, 1895, "What the Tortoise Said to Achilles," *Mind*, n. s. vol. 4, nº 14, pp. 278-280.

--- 1897, *Symbolic Logic, Part I: Elementary*, London: Macmillan (reprinted in *The Mathematical Recreations of Lewis Carroll: Symbolic Logic and The Game of Logic*, New York: Dover, 1958).

Corcoron, J., 1998, "Information-theoretic Logic", *in* C. Martinez, *et al.* (eds.), *Truth in Perspective*, Aldershot: Ashgate, pp. 113-135.

Englebretsen, G., 1974, "The tortoise, the turtle and deductive logic," *Jabberwocky*, vol. 3, pp. 11-13.

--- 1993-1994, "The properly Victorian tortoise," *Jabberwocky*, vol. 23, pp. 12-13.

--- 2008, "The Dodo and the DO: Lewis Carroll and the Dictum de Omni," *Proceedings of the Canadian Society for the History and Philosophy of Mathematics*, vol. 20, pp. 142-148.

--- 2010, "How to use a valid deriver's license", *The Reasoner*, vol. 4, pp. 54-55.

Mill, J. S., 1847, *A System of Logic*, New York: Harper & Bros.

Quine, W. V., 1976, *Ways of Paradox and Other Essays*, revised edition, Cambridge, MA: Harvard University Press.

--- 1981, *Theories and Things*, Cambridge, MA: Harvard University Press

Russell, B., 1903, *Principles of Mathematics*, Cambridge: Cambridge University Press.

Ryle, G., 1950, "If, so, and because", *in* M. Black (ed.) *Philosophical Analysis*, Ithaca: Cornell University Press.

Thomson, J. F., 1960, "What Achilles should have said to the Tortoise," *Ratio*, vol. 3, pp. 95-105.

Toulmin, S., 1950, *An Introduction to the Philosophy of Science*, London: Hutchinson.

Note 2021

This was the latest of my attempts to deal with that old Tortoise (and the Turtle). I returned to the question of what lesson about deductive logic was Carroll's title meant to convey. I rehearsed my attempts (and failures) to answer this question adequately in my 1974 and 1994. I also said a bit more about his contemporaries, J. S. Mill and F.H. Bradley, and how their worries about deduction were not his. This was followed by some remarks about how later philosophers dealt with the Tortoise's infinite regress argument that cast a shadow over the very possibility of deduction. Here are a few lessons I drew from Carroll after those years from 1974 to 2016. (1) Bradley was probably the one Carroll had in mind when he imagined the conversation the Tortoise had with Achilles. (2) Mill wasn't really a Turtle logician. (3) The difference between how Victorian logicians and modern logicians viewed deductive logic are significant. (4) Carroll wanted to defend deduction by defending the *dictum de omni*. (5) Such a defense requires a much better understanding of the old doctrine of distribution. (6) Carroll's challenge to logicians posed in this brief story is demanding and deceptively subtle.

17 Lewis Carroll's *Almost* Diagrammatic Logic Notation

with Amirouche Moktefi

Logic in Question, Jean-Yves Béziau, Jean-Pierre Descles, Amirouch Moktefi, and Anca Pascu (eds.), Basel :Birkhaüser, 2021

Abstract.

There is growing literature on the employment of diagrams to ease logical reasoning. Less is said about the visual properties of symbolic notations. It is, for instance, no coincidence that most notations offered for implication are asymmetrical to reflect the asymmetry of the operator itself. Hence, the symbol is self-interpreting in that its appearance suggests a property of the object it stands for. In this paper, we discuss a compositional notation, invented by Lewis Carroll in 1884, which suggests relations between propositions. Carroll is known for designing both symbolic and diagrammatic notations for logic. Although his theory was rooted in the 'old' logic, he championed a thorough use of notations in the Boolean style that was spreading in his time. In the notation we are considering here, Carroll represents simple propositions, then their symbols are combined to form compound propositions. An interesting outcome of this design, based on the visualisation of the composition of propositions, is that the layout exhibits relations between propositions. For instance, a proposition is shown to be subaltern to another if the symbol of the former is *contained* by the symbol of the latter. In this paper, we consider the guiding principles for the design of this notation and the extent to which it visually conveys relations between propositions.

Introduction

Lewis Carroll (Charles Dodgson) (1832-1898) carried out his logic investigations in that period between George Boole and Bertrand Russell. He was certainly an innovator in many of the areas of logic that interested him, but he was definitely no revolutionary (in any area of his life). He was a traditional logician. While he, like most of his contemporaries, was unaware of what Frege was setting in motion, he knew the ideas of Boole and other algebraic logicians [3, p. 6; 37]. Carroll shared with them, and with most British logicians at the time, a number of interests that have enjoyed relatively little attention or favour since. Among these are logical pedagogy, logical notation, and logical diagrams [36; 38]. Carroll was exceptionally innovative in all three areas (see [4]). In what follows, we make some observations about his ideas related to notation, with a particular interest in his first logic notation and the principles that guided its design.

1. It All Started with Aristotle

With the exception of a few mathematician-logician-philosophers, especially G. W. Leibniz, the 17th and 18th centuries displayed little sustained interest in (and often hostility toward) formal logic. Beginning in the 19th century, however, work on formal logic progressed rapidly. A number of factors contributed to this. Initially, and importantly, this was the result of the development of *algebraic logic* by a group of mid-century British mathematician-logicians, especially Boole [49]. By the end of the century, the new interest in formal logic resulted in a revolution of the entire logical enterprise, the outcome of which is today's standard version of formal logic – *mathematical logic*. A central figure of the revolution was Gottlob Frege. Pre-Fregean logic was, generally, a *term logic* initiated by Aristotle, whose version of term logic is usually referred to as *syllogistic*. Syllogistic was Leibniz's starting point. The same held for the British algebraic logicians. Frege rejected syllogistic (and term logic in general) and the resulting revolution was virtually total. Today's students of logic are encouraged to see traditional, pre-revolutionary logic as little more than a long series of historical curiosities and mistakes.

The problem of notation for formal logicians is to devise a symbolic system that (1) reveals the logical forms of statements made in a natural language (e.g., English, Greek, etc.), (2) while providing distinct kinds of symbols for the formal elements and the material elements of a statement, and (3) does so in a way that is consistent, perspicuous, relatively simple, and practical. So, in order to say what the logical form of a statement is, one first must distinguish between its formal and the material elements. The easiest thing to do here is to rely on the grammar of the language in which the statement is made. Plato and, at least before he developed syllogistic logic, Aristotle took the distinction found in their Greek between *onoma* and *rhema* (noun and verb) to reveal logical form. The theory here was that a *logos* (sentence, proposition, statement) was simply a combination of a noun and a verb; nothing more was required. But, of course, a sentence is not simply a list of words. What makes it a single unit rather than a list? The answer from Plato, (early) Aristotle, and (surprisingly) Frege is that the noun and verb somehow fit together to form one thing, a sentence. 'Man reasons' is a unified sentence, by Plato's lights, because the Forms, *MAN* and *REASON*, mix, mingle with one another. For Frege, the *onoma/rhema* distinction amounts to the *name/predicate* distinction. Names are "complete" expressions; predicates are incomplete (they have one or more gaps). A sentence is itself a complete, unified expression because the gaps in its predicate have all been filled by names.

In developing syllogistic logic, Aristotle saw that in the inferences that he wished to account for (syllogisms) there is always at least one expression that appears sometimes as a *subject term* and other times as a *predicate term*. Aristotle used the Greek word *horoi* limit, boundary, terminus) for *term*. He took a sentence to be a pair of terms bound together by a *logical copula*, thus forming a unit. Grammatical distinctions play no role in determining logical form. The latter is determined solely by the copulae, and there are four of them. In English, they can be expressed as: 'is said of every/all', 'is said of no', 'is said of some', and 'is not said of every/all'. The terms are the non-formal, the material elements. A copula comes between the two terms and joins, literally copulates, them. The terms are so-called because they are the expressions that are at the ends, limits, boundaries – the *termini*. Note that each copula indicates both a *quantity* (e.g., all, no, some) and a *quality* (is, is not). A sentence like 'Wisdom is said of some philosopher' is odd in English (just as its Greek version was as well). Medieval logicians, and subsequently traditional logicians generally, split the copula, attaching the quantity fragment to one term and the quality fragment to the other. The result was more natural (e.g., 'Some philosopher is wise'). The quantified term was called the *subject* and the qualified term was called the *predicate*. Thus, traditional logic is often called *subject-predicate logic*. It is important to keep in mind that the two fragments of the Aristotelian logical copula, while separated, nonetheless constitute a single formative element. Traditional logicians (perhaps following the lead of Abelard), and even modern logicians, call the positive qualifier (e.g., 'is') a copula. The logic of the algebraists, including Carroll, was such a subject-predicate logic.

Once a theory of logical form has been determined, the logician wants to devise a system of symbols that can readily display a sentence's logical form in a way that reveals the formal/material distinction and the quantity/quality distinction. Leibniz had hoped to produce a system of logical notation that would allow one to dispense with the relatively burdensome task of reasoning. Using the new formalized language, along with a system of rules for logical transformations of expressions in that language, a reasoner could simply *calculate*. Such a hope has been a key factor in driving the work of logicians in the 19th century and beyond. Algebraic logicians, such as Boole, Augustus De Morgan, John Venn, William S. Jevons, Carroll and Charles S. Peirce took seriously the challenge to contrive improved, practical notational systems (and even mechanical devices for making logical computations). Some, especially Venn, Carroll, and Peirce, invented graphic systems, diagrammatic languages meant to make logical forms, logical transformations and manipulations (e.g., deductions) visible. Of making many logic notations, there actually is no end.

2. Seeing Reason

Aristotle was the first systematic formal logician. In fact, he was also the first to use logical diagrams [53; 54]. Leibniz devised at least two different systems of logic diagrams, as did many others from the 17th to the 19th century [7; 29]. So the history of diagramming in logic is long; but it's also uneven (for a brief account of the history see [43], see also [22]). Euler and Venn diagrams are still recognized as efficient tools for exhibiting logical reasoning. Carroll developed a valuable variation on Venn diagram [11] (see [2; 34; 41]) and Peirce's system of diagrams, called Existential Graphs, more complex but more powerful, has gained increasing interest in recent years [45] (see also [47]). Post-Fregean logicians have tended to ignore or even disdain the use of diagrams in mathematics and logic. The assumption is that a diagram might be an aid for the novice but can be used in no other way [48]. Ironically, Frege's 2-dimensional system of logical notation [24] turns out to be *almost diagrammatic* [31; 32; 46].

An effective diagram is a kind of picture intended to resemble, in certain specified ways, what it depicts, represents. The relations among the parts of a diagram are analogous to those of the parts of what is represented. They are homomorphic. A good diagram intimately resembles what it represents, and what is represented can be concrete (in the way a good street map resembles the streets of a city) or abstract. In logical diagrams, including, Euler's, Venn's, and Peirce's, labeled figures such lines, circles, etc. are used to represent individual things or classes of individual things, and the relations among the individuals and classes is mirrored in the relations among the figures (see [35] for much more on this). Since resemblance is a matter of degrees (compare a photo of yourself to a stick-figure sketch), some diagrams depict their objects more closely than others. Euler diagrams are successful in this respect, others are less so.

One might think that resemblance is proper to diagrams and, thus, foreign to symbolic notations. That is far from true. Many notations were actually highly suggestive of the objects they represented. As such, we call them in this paper *almost* diagrammatic notations. This formulation is to be taken as a *bon mot* that conveys well our intended purpose, not as a literal statement: we do not mean that symbolic notations fall short of possessing diagrammatic features. On the contrary, we claim that many symbolic notations do give a visual hint of their meaning. By naming them *almost* diagrammatic, we merely mean that they are generally viewed as not having *enough* "diagrammaticity" to be called diagrams. Yet, they do have diagrammatic features as we shall see.

Boole, good algebraist that he was, construed sentences as equations. "Where Aristotle saw predications Boole saw equations" [13, p. 271]. Thus a sentence such as 'No man is a poet' can be paraphrased as 'Nothing is a man and poet', then as 'Man and poet is nothing', then symbolized as '(mp)=0'. Here the material terms, 'man' and 'poet' are symbolized by letters, juxtaposition indicates conjunction/intersection, quantity is indicated by '0', and quality by '=' (see [9]; for a critique of Boole see [12; 13]). De Morgan was, along with Boole, one of the most important and innovative 19th century British algebraic logicians. One of his innovations was the invention of a system of logical notation meant to replace Boole's. De Morgan referred to his as *spicular* notation (see [14; 16]; see also the important [26]). In one version of this notation, left and right parentheses are used to indicate quantity and quality and a dot to the left of left parenthesis indicates the negation of the term following the parenthesis (in another version the dot notation replaced by putting the term in lowercase). The four standard categoricals would be formulated as follows:

S))P	S).(P	S()P	S(.(P
A: Every S is P	E: No S is P	I: Some S is P	O: Some S is not P

Note that De Morgan's bundles of parentheses and dots operate, in effect, as unsplit copulae; 'S))P' could be read as 'P is said of every S'.

Spicular notation was flawed in some ways and never became widely used. Yet it did have an interesting characteristic (which De Morgan seems not to have exploited or even been aware of). The notion of distribution, perfected by medieval logicians, can be a powerful tool. Unfortunately, controversy has often surrounded the notion. Nonetheless, it can be useful in the analysis of syllogisms, providing easily applied rules for determining validity or invalidity. A subject term is said to be distributed in a sentence if it is universally quantified; undistributed if it is particularly quantified. A predicate term affirmed of a subject is undistributed; distributed if it is denied of a subject. Negation reverses distribution value (e.g., if 'not T' is distributed in a sentence, then 'T' itself is undistributed in that sentence). A syllogism is valid if and only if (1) the middle term is distributed at least once in the premises, (2) any term occurring in the conclusion has the same distribution value it had in the premises, and (3) the number of particular premises equals the number of particular conclusions. Obviously, any notation that clearly indicates terms' distribution values makes the job of determining syllogistic validity much easier. This is so of Carroll's underscoring technique which could indicate the distribution value of a term used in a given sentence (i.e., whether the term was distributed or undistributed). It holds for De Morgan's spicular notation as well (whether he realised this or not). Notice that any term immediately inside a parenthesis is distributed (the subject terms of A and E and the predicate terms of E and O; all other terms are undistributed).

3. A Simpler Notation than Boole's?

Carroll knew of Boole's equational logic and some other notations that were offered by subsequent algebraists to overcome the shortcomings of Boole's scheme. By the early 1880s, several notational systems were in competition among logicians [19; 39]. Christine Ladd-Franklin, who published her own scheme in 1883, reported that there were "in existence five algebras of logic, - those of Boole, Jevons, Schröder, McColl, and Peirce, - of which the later ones are all modifications, more or less slight, of that of Boole", and then proposed "to add one more to the number" [27, p. 17].

Carroll's attempts at a better notation than Boole's date back, at least, to 1876. On 25 May that year, he recorded in his journal expressing particular propositions such as "some x are y" in the form: "$xy > 0$" (instead of Boole's

"$vx = y$") [51, pp. 463-464] (the concatenation of the symbols xy indicate the aggregation of the terms that they stand for). It is known that Boole and his immediate followers had trouble representing particular propositions. Indeed, while it suffices to write "$xy = 0$" to indicate that "No x is y", one cannot write "$xy = 1$" to express the proposition "Some x are y" (because '1' stands for the universe). Booleans logicians sometimes introduce an additional sign (v for instance) to account for the nonempty intersection of x and y, so that $xy = v$. But mostly, they tended to overlook such propositions [42]. Carroll's solution ("$xy > 0$") is later found in Venn [50, p. 184-185]. In stark contrast to earlier attempt was the one Carroll seems to have referred on 20 November 1884:

> In these last few days I have been working at a Logical Algebra, and seem to be getting to a simpler notation than Boole's. [52, p. 153]

We are not told yet what this new notation (hereafter the 'first notation') looked like, but a subsequent entry, a month later, reveals it. Indeed, on 23 December 1884, Carroll says he has "given up [his] first notation" [52, p. 157]. This is his illustration of it:

E	"no x is y",	x ⌝ y,	or	xy ⌝
[E′]	"no x is not-y",	x ⌜ y,	or	x ⌝ y′
I	"some x is y",	x ∠ y,	or	xy ∠
O	"some x is not-y",	x ⦣ y,	or	x ∠ y′
A	"all x is y",	x ⌿ y		
[A′]	"all x is not-y"	x ⌿ y′		

Note that the prime sign: ' ′ ' stands for term negation (see [23]). This convention is already found in Carroll's *Euclid and his Modern Rivals* (1879) where we are told that an "accented letter denies what the letter asserts" [17, p. 24]. There are some interesting things to see in Carroll's first notation. In each case, the formative sign is placed between the two terms showing that it is, quite literally, a copula. In the *E* and *I* cases, however, an alternative configuration places the formative after the concatenation of the terms.

Whenever we are given a proposition asserting the relation between two (or more) terms, one may transform it into propositions of existence asserting or denying the joint presence of the terms. For instance, a proposition of relation "Some *x* is *y*" is transformed into a proposition of existence "Some *xy* exists". Likewise, a proposition "No *x* is *y*" is transformed into "No *xy* exists". This procedure, which was highly regarded by Carroll, was known to his predecessors, particularly Franz C. Brentano whose logical innovations were made known to British logicians by an account published in 1876 [28]. Carroll's first notation also features another practice, also found in Brentano, that consists in shifting the negation from the copula to the predicate. For instance, a proposition of the form "Some x is-not y" would be changed into "Some x is not-y", and eventually into "Some x not-y exists". This technique is applied in Carroll's first notation to propositions E′ and O for which two equivalent versions are given depending on whether it is the copula or the predicate that is denied.

It might be that the most intriguing feature of this notation is Carroll's claim that it is simpler than Boole's. Indeed, modern readers, accustomed to the mathematical costume of logic, will probably prefer the ingenious analogies suggested by Boole's equations. In order to make sense of Carroll's claim, we investigate in the next section what he viewed as the principles of notational design.

4. What Makes a Good Notation?

An early pamphlet of Carroll offers interesting insights on what he held to be a good notation. Indeed, in 1861, Carroll issued *The Formulae of Plane Trigonometry*, in which he attempted at a better notation for the trigonometric ratios (reprinted in [1, p. 121-139], see also [4; 5]). Carroll aimed to replace the "cumbersome expressions" used to indicate these ratios ('sin', 'cos', 'sec', 'tan', etc.) with better symbols. For the purpose, he enunciated what he viewed as the "four requisitions" of a good notation:

> [I] Setting aside, then, all alphabets, and (of course) all purely *mathematical* symbols, such as numerals, &c., no course remained but to *invent* new symbols for this purpose. In setting about this I took as principles – [II] to secure their being easily written, that each should consist of *two* strokes of the pen only – [III] to connect them with each

other, that *one* of these two strokes should be the same in all – [IV] to make them suggestive of their meaning, that they should represent (as nearly as possible) the *geometrical* lines to which these ratios belong. [1, p. 124] (the numbers are added)

One may contend that Carroll's first principle, to invent a new notation, contrasts with the fourth, that is the search for suggestiveness. Indeed, mathematicians customarily use the same symbols in various disciplines, especially when the adoption of existing symbols may suggest fruitful analogies. That was precisely the case among early Boolean logicians who adopted algebraic symbols to suggest analogies between the algebra of logic and 'quantitative' algebra. Venn explicitly lists "occasional *suggestiveness*" among the "advantages in employing an already widely-used and familiar set of symbols" [50, p. 104]. Yet, Venn warns us that "we must not permit such hints as these to be anything more than hints, for every logical rule must be established on its own proper grounds; but even hints may be of great value." [50, p. 105]. It is this risk that Carroll apparently was reluctant to take, as he preferred to introduce new notations in the mathematical areas that he explored, including logic (see [33]).

For the purpose of trigonometry, Carroll invites us to contemplate the angle AOB in the geometrical figure Fig. 1 Each of its trigonometric ratios can be easily discerned and described: BN is its *sine*, ON is *cosine*, TO is *secant* and AT its *tangent* [1, p. 123]. What Carroll noticed was that each ratio could be illustrated by a combination of the semi-circle and a line segment, where each ratio is uniquely represented by the position of its line segment relative to the semi-circle, as shown in Fig. 2. These symbols are unambiguous, simple, connected with one another, and, most importantly, suggestive of their meanings. Each is a kind of combination of a symbol and diagram – a simple diagram used as a symbol, a symbol looking like what it means.

Carroll's symbols may truly be said to be *almost* diagrammatic in that they give a visual hint of their meaning. Yet, despite its ingenious design, this notation did not circulate (see a review by De Morgan in [15]). This poor reception did not prevent Carroll from re-using it about thirty years later in a collection of mathematical problems [18, p. xix]. For our purpose, we will consider the extent to which he employed his four principles of notational design when he

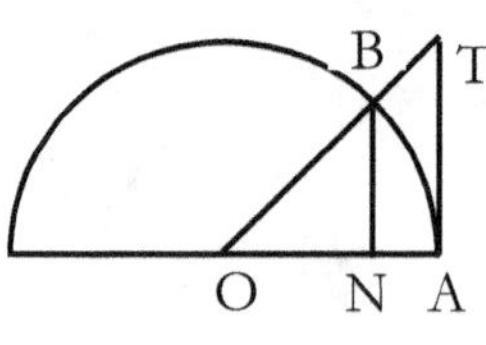

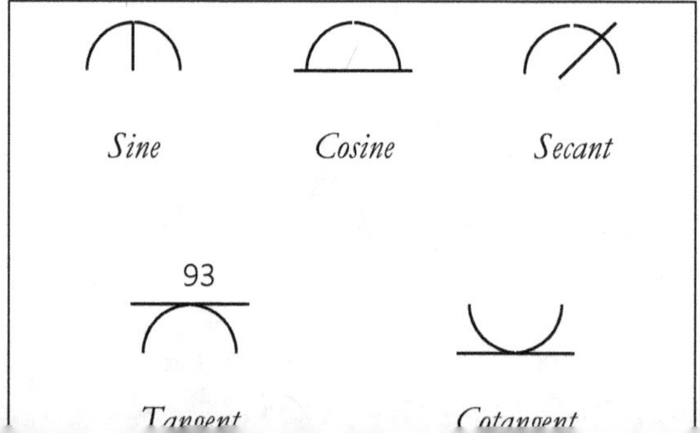

invented his first logic notation that he claimed to be simpler than Boole's.

5. Anatomy of a Notation

In the account of his notation, Carroll first expresses four simple propositions "no *x* is *y*", "no *x* is not-*y*", "some *x* is *y*" and "some *x* is not-*y*". These are found by considering how a term *x* relates to a term *y* and its negative not-*y*. To represent them, Carroll uses two strokes of pen: a horizontal line segment simply indicates that the two terms constitute a unified expression (Fig. 3*a*), then a slanted line segment, intersecting the horizontal, indicates which proposition is being represented (Fig. 3*b*-*e*). It is easy at this stage to see that Carroll's three first principles of notational design have carefully guided the conception of these symbols. First, he invented new symbols for his purpose rather than recycling existing notations. Second, the notation is simple as each symbol requires "*two* strokes of the pen only". And third, the horizontal stroke is the same in all the symbols, and hence, connects them to each other.

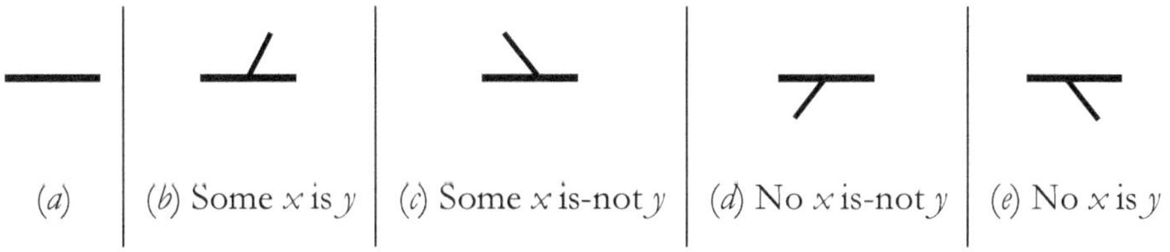

(*a*)	(*b*) Some *x* is *y*	(*c*) Some *x* is-not *y*	(*d*) No *x* is-not *y*	(*e*) No *x* is *y*

Fig. 3

Carroll's fourth principle (to make the notation suggestive of its meaning) is harder to detect, as the appearance of the symbols might at first look arbitrary. However, a closer look at the direction of the slanted segment, which indicates the quantity and quality of the propositions, reveals some regularities. Indeed, we can observe that the direction actually follows the pattern depicted in Fig. 4: Positive quantity (*Some*) is found above the line while the negative (*No*) is below, and positive quality (*Is*) is on the right while the negative is on the left (*Is not*). Note that the logical oppositions positive-negative are expressed by the familiar spatial relations up-down and right-left which convey well our intuitions of power constructs [30]. This scheme offers an easy mnemonic tool to memorise the meaning of the notation.

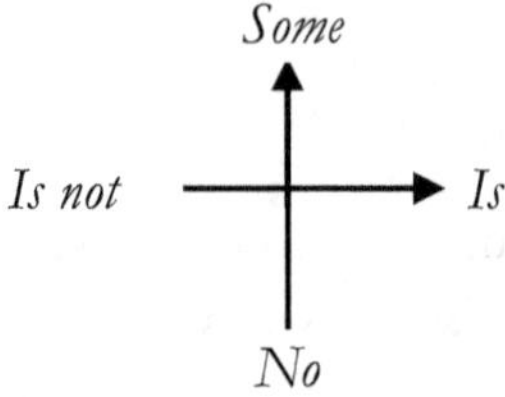

Fig. 4

Another advantage of this layout is that it allows symbols to suggest relations between the propositions. Let us recall how Carroll designed the *cotangent* symbol (⌣) in his trigonometric notation. He already had a tangent symbol (⌢) whose appearance (a short line segment literally tangent to an arc) readily suggested the *tangent* ratio (the thing to be represented). For the *cotangent* ratio, Carroll explained that he "simply *inverted* the [*Tangent*] symbol, which may be taken to indicate the fact that each is the *reciprocal* of the other […] which seems to be a very consistent and self-interpreting notation" [1, p. 125]. A similar process is at work in the logic notation where propositions which differ in quality or quality simply have their symbols reversed to suggest their opposition.

6. Composition and Suggestiveness

So far, we discussed only simple propositions. We now turn to compound propositions which reveal a powerful feature of Carroll's notation and their suggestiveness. In the 1880s, Carroll commonly conceived simple proposition, as explained above, then combined them to form complex ones. For instance, on 23 September 1885, he recorded in his journal completing "the interlacing triangle of 13 propositions, involving x as subject, with or without z or z´ as predicate, exhibiting their composition" [52, p. 239]. This line of research led Carroll to design several charts that recently got attention in schoarship on logical oppositions [44].

Propositions of the form "All x are y" offer a simple example of such compositions since Carroll viewed them as double: they are formed by the combination of "Some x is y" and "no x is not y". Now, the interesting fact is that, in Carroll's notation, the symbol of the compound proposition is truly the combination of the symbols of the propositions it combines:

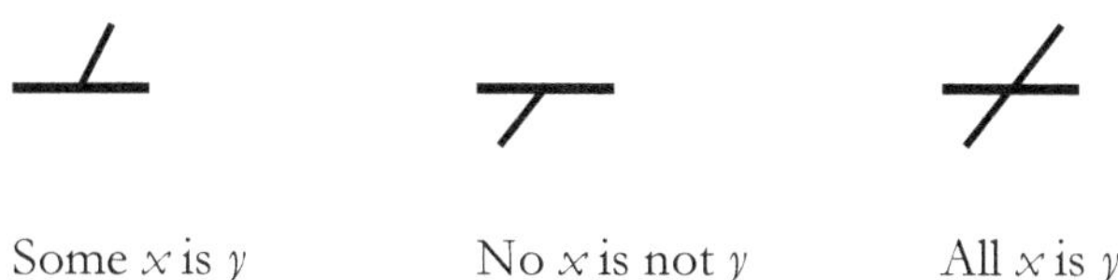

In this sense, the new symbol is self-interpreting because it *has* the relation it is intended to convey. Unfortunately, the form "All *x* are *y*" was the only compound form that Carroll expressed with his first notation. Hence, it is unknown whether he generalised this technique to all compound propositions. Yet, it does not take much to see how fertile this approach may be. Let us consider, for instance, the "four *conceivable* states of things" between a class *x* and two classes *y* and not-*y*, that Carroll listed in his *Symbolic Logic*: (1) "Some *x* are *y*, and Some are not-*y*"; (2) "Some *x* are y, and none are not-*y*"; (3) "No *x* are *y*, and some are not-*y*"; and (4) "No *x* are *y*, and none are not-*y*" [11, p. 195]. Carroll noted that form (2) is equivalent to "All *x* are *y*", (3) to "All *x* are not-*y*", and (4) "No *x* exist". Somehow, he did not provide a simple equivalent expression to form (1), even though he recorded in his diary, on 30 March 1887, that it was equivalent to the "fancy" proposition "only some *x* are *y*."'" [52, p. 327]. Since each "state" is the combination of two simple propositions that are simply expressed with Carroll's first notation, the expression of the compound propositions easily follows:

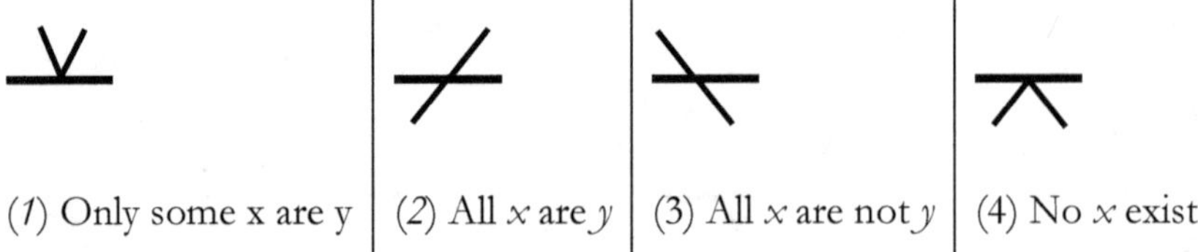

It is possible to get information about the logical relation between given propositions by a simple look at the appearance of their symbols. Subalternation is the simplest instance: a proposition is subaltern to another if the symbol of the former is "contained" by the symbol of the latter. Hence, in the following, proposition (*b*) is subaltern to (*a*). Simple propositions (*c*) and (*d*) are complementary, a fact that is depicted by their opposite slanted segments. Propositions (*e*) and (*f*) are incompatible because they *contain together* complementary propositions. Finally, each of propositions (*g*) and (*h*) is inconsistent because *each contains* incompatible propositions.

(*a*) (*b*)	(*c*) (*d*)	(*e*) (*f*)	(*g*) (*h*)
Subalternation	Complementarity	Incompatibility	Inconsistency

The beauty of this notation springs from its visualisation of the composition of compound propositions. As such, it permits the discovery of relations between propositions by the mere contemplation of their symbols. In this sense, it may be said that the very appearance of a given symbol (or set of symbols) resembles the thing being represented closely enough to suggest immediately to the reader the thing itself

7. There is Always Better

Despite its beauty and suggestiveness, Carroll apparently abandoned his first notation just one month after its conception. It is unclear what motivated this hasty decision, but it seems that he did not find them appropriate to tackle disjunctive propositions. Indeed, in December 1884, Carroll introduced to his system propositions of the form "Not all *x* are y" which are equivalent to the disjunctive "No *x* exist, or, Some *x* are not *y*" [52, p. 158]. Such forms are evidently difficult to represent. Conjunctions are easier because it is intuitively assumed that the co-location of symbols accounts for their co-existence. If one draws a cup and a plate on a table, it is presumed that *both* objects are on the table. If one wishes to state that *at least one* object is on the table, an additional device is needed to convey this meaning. This feature made disjunctives difficult to express and it cannot be said that Carroll's subsequent attempts were more successful.

Yet, it is interesting to survey some of Carroll's successive notations in order to identify some of his notational concerns. We consider here their evolution through the lens of his representations of the *A* proposition "All *x* are *y*" and the extent to which they conveyed its composition:

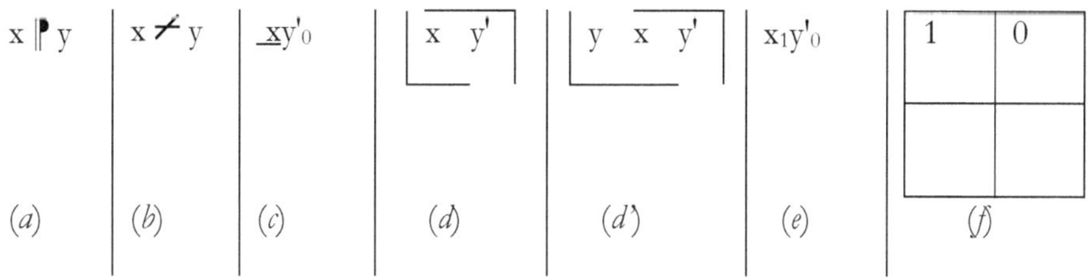

Symbol '⁋', in (*a*), was introduced in 1879 and, since, regularly used in a sense that roughly corresponds to logical implication [40]. But in some manuscripts, Carroll used it also to represent the copula so that "*x* ⁋ *y*" would express "All *x* are *y*" [6, pp. 256-257]. The system of notation (*b*), which is the object of our paper, was designed in November 1884 and directly depicts the decomposition of "All *x* are *y*" into "Some x are y" and "No x is not y". Then, in December 1884, Carroll preferred a new notation (*c*) where existent terms are underscored while empty ones are marked with a subscript '0' [52, p. 158]. Note that it is *x* alone that is underscored; hence the proposition "All *x* are *y*" should be read as the composition of "Some *x* exists" and "No *x* is *y'*". Carroll later justified this variation, arguing that "the Proposition "Some *xy* exist" contains *superfluous information*. "Some *x* exist" is enough for our purpose" [11, p. 72]. Indeed, if some *x* exists, and given that no *x* is not-*y*, it follows that some *x* are *y* (see [20, pp. 40-41]). In September 1885, Carroll recorded again devising "better symbols" with notation (*d*). Here, a system of lines is generalized: underscoring indicates existence and overscoring emptiness. Again, the subordination of *I* propositions to *A*s is not directly depicted but Carroll noted that one may use another form (*d'*) which "exhibits fact that "A" contains I"" [52, p. 239].

Then, starting from 1886, Carroll settles for a generalisation of subscripts: '1' indicates existence and '0' for emptiness. This notation (*e*) was eventually used in *Symbolic Logic* (1896) where underscoring was reserved to indicate the distribution of terms [11] (see [21]). The proposition "All *x* are *y*" is still shown as the composition of "Some *x* exists" (x_1) and "No *x* is *y* " (xy'_0). It is possible to convey the subalternation of *I* to *A* by providing a decomposed expression of the latter in the form "xy_1 † xy'_0", where '†' stands for sentential conjunction. Here, the simple propositions that compose *A* are connected but not merged. This technique obviously can be used with each of Carroll's previous notations, but such expressions are less diagrammatic than those, we considered so far, where the merging of the simple propositions readily produces the compound proposition without additional syntactic devices.

This survey shows that, among Carroll's symbolic notation, the first one alone directly suggest the subalternation of *I* to *A* propositions. Interestingly, it shares this distinction only with Carroll's diagrams (*f*), where Carroll simply marks a cell with '1' if it exists and '0' if it does not. Hence, the expression of "All *x* are *y*" is done by placing '1' on the *xy* cell and '0' on the *x not-y* cell. Intriguingly, Carroll invented his diagrams in 1884, just one week after the conception of his first notation [52, p. 155], even though their layout has subsequently changed before their publication in 1886 [10]. Naturally, Carroll's other notations had their own merits and are not deprived of suggestiveness. The use of subscripts '1' and '0', for instance, were chosen to suggest that "there is at least one Thing" and "there is no Thing", respectively [11, p. 176]. Most importantly, except for (*a*), all Carroll's notations exhibit the composition of complex propositions and, thus, are suggestive of their meaning.

8. A Diagrammatic Variation

Carroll's first notation exhibited explicitly and adequately his acceptance of the existential import of universal affirmations and the subordination of particular affirmations. It may be worthwhile considering what the notation would have looked like if these assumptions were released. In the following, we consider some variations on Carroll's first notation, including a diagrammatic method which it inspired.

Were it not for Carroll's insistence on explicit existential import for universal affirmations, the four categoricals could simply be formulated as shown in the variation (1). The change consists in simply making the universal affirmation "All x is y" equivalent to "No x is not-y". In this case, the four categoricals correspond simply to Carroll's four primitive propositions, prior to compositions. We may introduce negated predicates to ease the notation, as shown in the variation (2). This scheme offers a clear distinction between universals and particulars, while quality is simply ascribed to the predicate. In variations (1) and (2), the universal affirmation (A) is transformed into a negation (E) and transcribed accordingly. Another possibility consists in transforming the universal negation into an affirmation: "No x is y" simply becomes "All x is not-y" and is transcribed accordingly, using Carroll's original (composed) symbol, as shown in variation (3). This new scheme has remarkable features. Indeed, all its forms are affirmations: each categorical is either a universal affirmation or a particular affirmation and, in each case, the predicate term may or may not be negative. As such, it certainly achieves a significant degree of simplicity while paying the small price of introducing negative terms, something Carroll was generally quite willing to do.

	Original	Variation (1)	Variation (2)	Variation (3)
A	$x \not- y$	$x \overline{/} y$	$x \overline{/} y$	$x \not- y$
E	$x \overline{\smallsetminus} y$	$x \overline{\smallsetminus} y$	$x \overline{/} y'$	$x \not- y'$
I	$x \angle y$	$x \angle y$	$x \angle y$	$x \angle y$
O	$x \underline{\smallsetminus} y$	$x \underline{\smallsetminus} y$	$x \angle y'$	$x \angle y'$

This variation (3) is suggestive enough to be the foundation of a full diagrammatic system for syllogistic. Indeed, while Carroll's symbols for trigonometric relations are almost diagrammatic, this streamlined version of his logical notation *is* diagrammatic. The now-streamlined version of Carroll's logic notation is not on par with Euler's. Nevertheless, when relatively simple but appropriate conventions and rules for constructing, interpreting, and manipulating such diagrams are provided, it can still be an effective tool for use in tasks of logical reckoning. The modified Carroll notation amounts to just such a system for syllogistic. For a given syllogism, each premise is formulated in the new Carroll symbols and then each such formula is used as a diagram for drawing the appropriate conclusion. Here an important rule is that the conclusion follows the weakest premise (the medieval logicians referred to this as the *peiorem* rule). Thus, if one premise is particular then so is the conclusion; if one premise is negative then so is the conclusion. In each case, the premises are combined to form a single diagram, the middle term, y, is eliminated (see [25]); and the result is the conclusion. Now, given the possibility that any predicate term might be negative, every categorical says, in effect, that the class represented by the subject term is completely contained (for universals) or partially contained (for particulars) in the class represented by the predicate term. Indeed, the horizontal line of predication can be read as 'is … contained in' and the intersecting line can be read as indicating quantity, 'completely' or 'partially', filling the lacuna.

Let us illustrate this procedure by treating each of the four perfect first figure syllogisms, starting with *Barbara.* After representing each premise separately (Step 1), we merge their figures with the middle term y connecting the two other terms x and z, each term being in this case affirmed (Step 2).

Then we eliminate the middle term, y, and we get the diagram of the conclusion (Step 3). Its interpretation provides the final conclusion of the syllogism (Step 4)

Barbara

"All *x* is *y*" and "All *y* is *z*"

1. *Representation*
2. *Conjunction*
3. *Elimination*
4. *Interpretation*

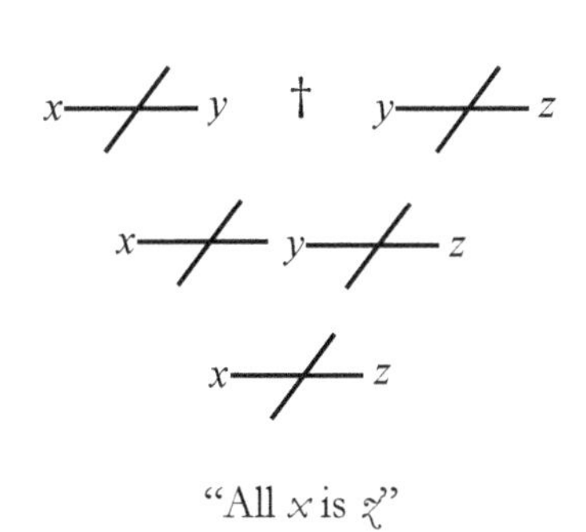

A similar analysis easily conveys how the conclusion of other syllogisms follows from their pairs of premises. Here is shown the analysis of other perfect first figure syllogisms:

1.
2.
3.
4.

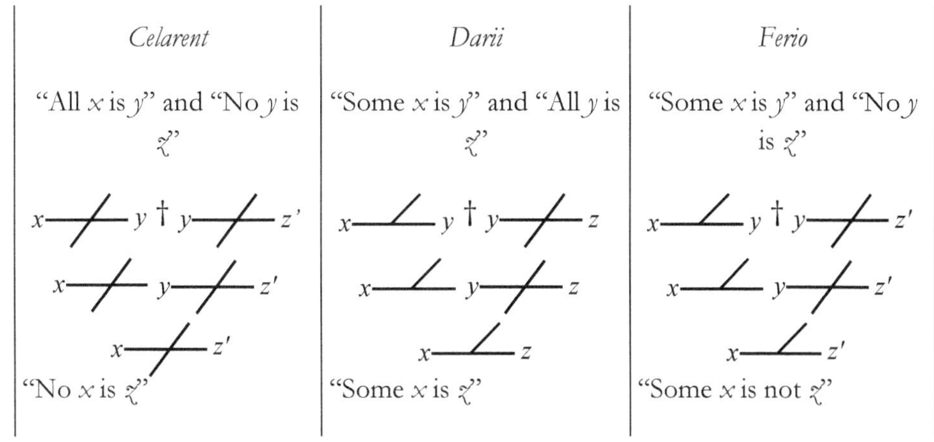

In each of these cases, a merging diagram is derived by Conjunction from the preceding diagram pair and the final diagram is then derived from it by Elimination. It suffices to interpret the final diagram to assert the conclusion. The main features of this scheme are: (1) the transformation of negations into affirmations, which makes the diagrams consist in a sequence of (complete or partial) inclusions. This technique was known to past logicians and is found in Peirce's amendment of Euler diagrams to handle negative terms [8]. (2) the iconic representation of the 'quantity' of inclusion by the magnitude of the slanted segment: a complete segment (on both sides of the horizontal segment) expresses complete inclusion while a partial segment (on one side of the horizontal segment) expresses partial inclusion.

Naturally, this technique can be applied to more complex problems, such as sorites, provided that the series of propositions is conveniently arranged. A sorites such as "All *x* is *y*; No *y* is *z*; Some *z* is *w*; thus, Some *w* is not *x*" would have a similar analysis. Its diagram would simply be:

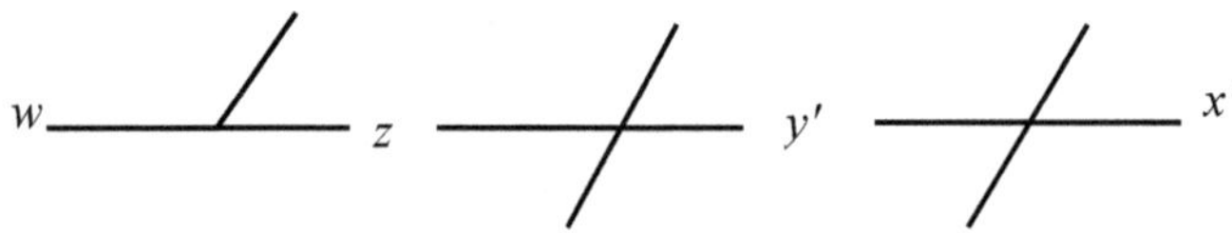

It is easy to read the conclusion "Some *w* is not *x*" after eliminating middle terms *z* and *y'*.

Conclusion

Near the close of 1884, Lewis Carroll confided to his diary a suggestion for a new system of symbolic notation for logic. He was in the process of looking for an improvement on Boole's logical notation. Shortly afterwards, he settled on a modification of that Boolean system. Yet his earlier suggested notation was completely original, with a number of features in its favour. One of these features was a result of the *appearance* of his symbols – they seemed almost diagrammatic. In fact, though there are better, more powerful diagrammatic systems for logic, a workable system of diagrams, at least for syllogistic, can be extracted from Carroll's suggested notation. He never did that. Within a few short years, he had not only turned to favour his own version of Boole's *symbolic* system but had developed his own (impressive) version of Venn's *diagrammatic* system as well.

Acknowledgements

In addition to expressing our appreciation to Miles Rind for his help and advice, we want to thank Francine Abeles for her valuable suggestions and critique. Fran has provided us with so many of the fruits of her research over many years, answering our questions while making astute suggestions with generosity and patience. For the second author, this research benefitted from the support of ERC project 'Abduction in the age of Uncertainty' (PUT 1305, Principal Investigator: Prof. Ahti-Veikko Pietarinen).

References

Abeles, F. F., ed.: *The Mathematical Pamphlets of Charles Lutwidge Dodgson and Related Pieces*. Lewis Carroll Society of North America, New York (1994).

Abeles, F. F.: Lewis Carroll's visual logic. History and Philosophy of Logic **28**, 1-17 (2007).

Abeles, F. F., ed.: *The Logic Pamphlets of Charles Lutwidge Dodgson and Related Pieces*. Lewis Carroll Society of North America, New York (2010).

Abeles, F. F.: Mathematical legacy. In: Wilson, R., Moktefi, A. (eds) *The Mathematical World of Charles L. Dodgson (Lewis Carroll)* (pp. 177-215). Oxford University Press, Oxford (2019).

Abeles, F. F.: Charles L. Dodgson's work on trigonometry. *Acta Baltica Historiae et Philosophiae Scientiarum* **7**, 27-38 (2019).

Bartley III, W. W., ed.: *Lewis Carroll's Symbolic Logic*. Clarkson N. Potter, New York (1986).

Bellucci, F., Moktefi, A., Pietarinen, A.-V.: Diagrammatic autarchy. Linear diagrams in the 17th and 18th centuries. In: Burton, J., Choudhury, L. (eds.), *DLAC 2013: Diagrams, Logic and Cognition* (pp. 23-30). CEUR Workshop Proceedings, vol. 1132. (2014).

Bhattacharjee, R., Moktefi, A., Pietarinen, A.-V.: The representation of negative terms with Euler diagrams. In: Béziau, J.-Y., et al. (eds) *Logic in Question.* Birkhaüser, Basel (2020).

Boole, G.: An Investigation of the Laws of Thought. Macmillan, London (1854).

Carroll, L.: *The Game of Logic.* Macmillan, London (1886).

Carroll, L.: *Symbolic logic: Part I.* Macmillan, London (1897).

Corcoran, J.: Introduction. *The Laws of Thought: George Boole.* Prometheus Books, Amherst, NY. (2003).

Corcoran, J.: Aristotle's *Prior Analytics* and Boole's *Laws of Thought. History and Philosophy of Logic* **24**, 261-288 (2003).

De Morgan, A.: *Formal Logic.* Taylor & Walton, London (1847).

De Morgan, A.: Review of C. L. Dodgson's *The Formulae of Plane Trigonometry. The Athenaeum* **1761**, 113 (27 July 1861).

De Morgan, A.: *On the Syllogism and Other Logical Writings.* Routledge & Kegan Paul, London (1966).

Dodgson, C. L.: *Euclid and his Modern Rivals.* Macmillan, London (1879)

Dodgson, C. L.: *Pillow Problems.* Macmillan, London (1895).

Durand-Richard, M.-J., Moktefi, A.: Algèbre et logique symboliques : arbitraire du signe et langage formel. In : Béziau, J.-Y. (ed.) *La Pointure du Symbole* (pp. 295-328). Pétra, Paris (2014).

Englebretsen, G.: Lewis Carroll on logical quantity. *Jabberwocky* **12** (2), 39-41 (1983).

Englebretsen, G.: The Dodo and the DO: Lewis Carroll on the Dictum de Omni. *The Carrollian* 25, 29-37 (2014).

Englebretsen, G.: *Figuring it Out.* De Gruyter, Berlin (2020).

Englebretsen, G., Gilday, N.: Lewis Carroll and the logic of negation, *Jabberwocky* **5** (2), 42-45 (1976).

Frege, G.: *Begriffsschrift.* Louis Nebert, Halle (1879).

Green, J.: The problem of elimination in the algebra of logic. In: Drucker, T. (ed.) *Perspectives on the History of Mathematical Logic* (pp. 1-9), Birkhauser, Boston (1991).

Heinemann, A.-S.: 'Horrent with mysterious spiculae': Augustus De Morgan's logic notation of a 'calculus of opposite relations'. *History and Philosophy of Logic* **39**, 29-52 (2018).

Ladd-Franklin, C.: On the algebra of logic. In: Peirce, C. S. (ed) *Studies in Logic* (pp. 17-71). Little, Brown, and Company, Boston (1883).

Land, J. P. N.: Brentano's logical innovations. *Mind* **1** (2), 289-292 (1876).

Lemanski, J.: Periods in the use of Euler-type diagrams. *Acta Baltica Historiae et Philosophiae Scientiarum* **5** (1), 50–69 (2017).

Lu, A., *et al.*: What directions do we look at power from? Up-down, left-right, and front-back. *PLoS ONE* 10(7): e0132756. doi:10.1371/journal.pone.0132756, pp. 1-16

Macbeth, D.: Diagrammatic reasoning in Frege's *Begriffsschrift. Synthese* **186**, 289-314 (2012).

Macbeth, D.: Writing Reason. *Logique et Analyse* **221**, 25-44 (2013).

Marion, M., Moktefi, A.: La logique symbolique en débat à Oxford à la fin du XIXe siècle : les disputes logiques de Lewis Carroll et John Cook Wilson. *Revue d'Histoire des Sciences* **67** (2), 185-205 (2014).

Moktefi, A.: Beyond syllogisms: Carroll's (marked) quadriliteral diagram. In: Moktefi, A., Shin, S.-J. (eds.) *Visual Reasoning with Diagrams* (pp. 55–72). Birkhäuser, Basel, (2013).

Moktefi A.: Is Euler's Circle a Symbol or an Icon?. *Sign Systems Studies* **43** (4), 597–615 (2015).

Moktefi, A.: On the social utility of symbolic logic: Lewis Carroll against 'The Logicians'. *Studia Metodologiczne* **35**, 133-150 (2015).

Moktefi, A.: Are other people's books difficult to read? The logic books in Lewis Carroll's private library. *Acta Baltica Historiae et Philosophiae Scientiarum* **5** (1), 28-49 (2017).

Moktefi, A.: Logic. In: Wilson, R. J., Moktefi, A. (eds.) *The Mathematical World of Charles L. Dodgson (Lewis Carroll)* (pp. 87-119). Oxford University Press, Oxford (2019).

Moktefi, A.: The Social shaping of modern logic. In: Gabbay, D., *et al.* (eds) *Natural Arguments: A Tribute to John Woods* (pp. 503-520). College Publications, London (2019).

Moktefi, A., Abeles, F. F.: The making of 'What the Tortoise said to Achilles': Lewis Carroll's logical investigations toward a workable theory of hypotheticals. *The Carrollian* **28**, 14-47 (2016).

Moktefi, A., Edwards, A. W. F.: One more class: Martin Gardner and logic diagrams. In: Burstein, M. (ed.) *A Bouquet for the Gardener* (pp. 160-174). The Lewis Carroll Society of North America, New York (2011).

Moktefi, A., Pietarinen, A.-V.: On the diagrammatic representation of existential statements with Venn diagrams. *Journal of Logic, Language and Information* **24** (4), 361–374 (2015).

Moktefi, A., Shin, S.-J.: A history of logic diagrams. In: Gabbay, D. M., Pelletier, F. J., Woods, J. (eds) *Logic: A History of its Central Concepts* (pp. 611-682), North-Holland, Amsterdam (2012).

Moretti, A.: Was Lewis Carroll an amazing oppositional geometer?. *History and Philosophy of Logic* **35** (4), 383-409 (2014).

Peirce, C. S., Pietarinen, A.-V.: *Logic of the Future: Writings on Existential Graphs.* Vol. 1, De Gruyter, Mouton (2019).

Schlimm, D.: On Frege's *Begriffsschrift* notation for propositional logic. *History and Philosophy of Logic* **39** (1), 53-79 (2018).

Shin, S.-J.: *The Iconic Logic of Peirce's Graphs.* MIT Press, Cambridge, MA (2002).

Tennant, N.: The withering away of formal semantics?. *Mind & Language* 1, 302-318 (1986).

Van Evra, J.: The development of logic as reflected in the fate of the syllogism 1600-1900. *History and Philosophy of Logic* **21**, 115-134 (2000).

Venn, J.: *Symbolic Logic.* Macmillan, London (1894).

Wakeling, E., ed.: *Lewis Carroll's Diaries.* vol. 6, The Lewis Carroll Society, Clifford, Herefordshire (2001).

Wakeling, E., ed.: *Lewis Carroll's Diaries.* vol. 8, The Lewis Carroll Society, Clifford, Herefordshire (2004).

Wesoły, M.: Aristotle's lost diagrams of the analytical figures. *Eos* **84**, 53-64 (1996).

Wesoły, M.: ΑΝΑΛΥΣΙΣ ΠΕΡΙ ΤΑ ΣΧΗΜΑΤΑ: Restoring Aristotle's lost diagrams of the syllogistic figures. *Peitho* **3**, 83-114 (2012).

18 Lewis Carroll on Some Fundamental Concepts in Logic

South American Journal of Logic, 2021, to appear

1. "A Very Subtle Difficulty"

In later editions of *Symbolic Logic, Part I*, Carroll introduced (in Book II, Chapter II, Propositions of Existence) a new account of the logic of *existing things* that had generated a "very subtle difficulty" in the first edition. In 1957, reviewing *Symbolic Logic*, the philosopher-logician Arthur Prior (1914-1969) appears to be the first to have noted the import of such a change. He wrote:

> [I]n earlier editions Carroll gives as the "normal forms" (his own phrase) of his two sorts of existential propositions "Some S's exist" and "No S's exist." Now, however, the normal forms are "Some existing things are P" and "No existing things are P's." By this alteration, he claims, he is "evading a very subtle difficulty which besets the other form"; meaning no doubt that the first form suggests, as the second does not, that S's are in principle divisible into ones that exist and ones that do not. (Prior 1957, 310)

It seems that Carroll saw what he took to be a good reason for changing his mind about propositions such as 'Some honest men exist'. Having first analyzed this as (1) 'Some honest men are existing things', he then found a "very subtle difficulty" requiring a new analysis: (2) 'Some existing things are honest men'. "Subtle" indeed. One's initial reaction would be that the two versions are logically equivalent (by simple conversion). Prior saw that any proposition of the form 'Some X is ..' can be construed as either referring to the X things that exist (= are in the universe of discourse) or to the X things that exist or don't exist. Carroll took this to be the problem of saying what a thing is, where 'thing' is understood as 'in the universe of discourse'. So now consider again the two analyses. The subject of (1) is 'Some honest men'; the subject of (2) is 'Some existing things'. (2), unlike (1), wears its existential status on its face. The subject of (1) amounts to 'Some things that are honest men' (for Carroll) or 'Some existing honest men' (for Prior). In either case, the resulting analysis would be (1c) 'Some things that are honest men are honest men' or (1p) 'Some existing honest men are existing things'. Carroll was right to make the change. While subject-terms and predicate-terms might exchange their roles in a proposition (e.g., by application of conversion), subjects and predicates can't do so.

2. "The Bewildering Question"

In the prefaces to the second, third, and fourth editions of *Symbolic Logic*, Carroll ended his brief preview of the change he made from the first edition concerning his evasion of his "very subtle difficulty" with reference to such "subtle difficulties that seem to lie at the root of every Tree of Knowledge." He added that "they are *far* more difficult to grapple with than any that occur in its higher branches." Carroll concluded that even the most foundational of difficulties in geometry are "'trifles, light as air,' compared with the bewildering question 'What is a Thing?'" So, what did he think a Thing is and how did he come to change his mind?

Carroll had noted in the prefaces to these later editions that he had adopted an alternative definition of *Classification.* He wrote that "this enabled me to regard the whole Universe as a 'Class', and thus dispense with the very awkward phrase 'a Set of Things'." This matter is taken up in Book I, Chapter II. Classification is said to be a process of imagining selecting a certain "Set of Things, all that have a certain Attribute (or Set of Attributes)" (first edition), "Things" (later editions), the result of which is a **Class**. It's unclear why he thought the phrase 'a Set of Things' was awkward. After all, in later editions he made use of the phrase 'the Class of "Things"', which seems more awkward. Nonetheless, *Things* are prominent in accounting for *Classification* and *Classes*. And they are the first thing Carroll gets to at the very beginning of Book I, Chapter I, Introductory, where he get right to the point: "The Universe contains **Things**." Well, what *is* a Thing? Carroll came to the view that classification is a mental process that can be performed independently of whether the Things being classified exist (are real) or not (are imaginary). This notion had a profound effect on how Carroll treated the problem of existential import in formal logic. In particular, it meant that universally quantified statements of the form 'Every S is P', 'All A are B', 'No P are Q', etc., if they are to be taken as referring to existing things, must be accompanied by an indication of this (e.g., 'Every (existing/real) S is P' or 'Every S is P and something is S'.

It must also be noted here that Carroll, along with his contemporary algebraic logicians, took seriously the idea that statements are always made relative to what was usually called a *universe of discourse.* A universe of discourse is a totality of things and is determined by the context of the statement itself. For Carroll, the universe of discourse could be his "Universe of Things" consisting of both things that exist and things that do not. However, a statement could be made in a context that implicitly assumes a universe of discourse consisting of just existing things, or existing red things, or red planets, or unreal things, or things found only in Wonderland.

3. "A Name of That Thing"

In Book I, Chapter IV, Names, Carroll considered *names* as singular terms or expressions such as proper names ('Plato') as well as what are now called definite descriptions ('the teacher of Aristotle'). As expected, real names "represent" existing things; unreal names represent things that do not exist. In Chapter II he had said that a single thing is a class consisting of just that one thing, an **Individual**. When he came to discussing propositions (Book II), Carroll said that a singular proposition, a proposition with an individual subject, must be taken as a universal statement, one with an implicit universal quantifier (e.g., 'every', 'all'): "A Proposition, whose Subject is an *Individual*, is to be regarded as *Universal*." (Bartley 68) As it happens, this was a policy that had been widely accepted by medieval logicians. Carroll's example was 'John is not well'. His original claim about such a proposition was that it should be properly construed as 'All Johns are men who are not well'. Here, the subject 'Johns' represents the class of men named 'John'. However, he soon came to change his mind (see Abeles and Moktefi 2011). So, in the next two editions, he took the subject term 'John' to represent only the individual (i.e., the one-member class consisting of) John. Finally, in the fourth edition, Carroll sought to clarify his understanding of such singular propositions, by writing that 'John' in his example represents the class of men referred to by the speaker when using that name. Thus, the proposition now becomes: "*All* the men, who are referred to by the speaker when he mentions 'John', are not well." (Bartley 1986, 68-69) Carroll had been faced here with a logical difficulty that he couldn't easily deal with. The difficulty was due in large measure to his failure, pointed out by Abeles and Moktefi, draw a distinction between what he called a *class* and what logicians now call a *set.* The latter is understood as an abstract object; it is determined by its members but is not identical with its member(s). By contrast, Carroll's class is nothing more than its constituent(s).

Unfortunately, Carroll's logical difficulties regarding singular subject terms did not end there. It's doubtful that he ever got fully clear about the logical syntax of propositions with such subjects. In particular, he seems to have held the old medieval notion of construing these as universally quantified without considering *why*. One reason why traditional logicians had treated singular propositions as universals was that it helped account for the fact that the subject terms of singulars were like those of universals in being *distributed*. Two centuries before Carroll faced the logical difficulty of determining the appropriate treatment of singulars, Leibniz had formulated a solution, one that went beyond the scholastics' *ad hoc* fix. He did so in a single paragraph:

> Some logical difficulties worth solution have occurred to me. How is it that opposition is valid in the case of singular propositions – e.g. 'The Apostle Peter is a soldier' and 'The Apostle Peter is not a soldier' – since elsewhere a universal affirmative and a particular negative are opposed? Should we say that a singular proposition is equivalent to a particular and to a universal proposition? Yes, we should. So l when it is objected that a singular proposition is equivalent to a particular proposition, since the conclusion in the third figure must be particular, and can nevertheless be singular; e.g. 'Every writer is a man, some writer is the Apostle Peter, therefore the Apostle Peter is a man'. I reply that here also the conclusion is really particular, and it is as if we had drawn the conclusion 'Some Apostle Peter is a man'. For 'some Apostle Peter' and 'every Apostle Peter' coincide, since the term is singular. (Leibniz 1966, 115)

Now, Leibniz accepted, as did all traditional logicians, including Carroll, that singular propositions must be taken as having at least some implicit quantity. What Leibniz meant when he wrote that 'some Apostle Peter' and 'every Apostle Peter' coincide was that, given the presupposition that there is just one Apostle Peter, reference to Apostle Peter was indifferent as to which quantity that might be – because, there being only one such referent, one Carrolian Individual, reference to every Apostle Peter amounts to a reference to at least one Apostle (i.e., to the one, and only, Apostle Peter. Leibniz's solution is semantic, depending on the notion of reference. In such cases as singular propositions, the logical syntax follows. Knowing that some man is a writer, the conclusion that every man is a writer does not immediately follow. However, knowing that (some) Shakespeare is a writer, it follows (not formally, but materially, via the understanding that there is but one Shakespeare) that (every)

Shakespeare is a writer. In ordinary discourse, we ignore (indeed, are usually ignorant of) any quantifier here because it makes no difference. As Leibniz said, the two expressions *coincide.*

Traditional logic is a *term logic* (in contrast with modern mathematical logic, which is generally said to be a first-order *predicate logic*). Nonetheless, there are now newer, modern versions of term-logic. In the case of the most prominent one, Leibniz's idea of taking the implicit quantity of singular propositions to be indifferently particular or universal is known as *Leibniz's wild quantity thesis*. For more on wild quantity see: Sommers 1967, 1969, 1970, 1976a, 1976b, and pp. 15-17 in 1982; Englebretsen 1986, 1988, and pp. 60-64 in 1996; Sommers and Englebretsen 2000, pp. 73-76.

4. "A Puzzling Question"

Carroll held, in his chapter on Classification (*Symbolic Logic*), that one could form the class of all Things (i.e., the Universe), or one could a class of those Things in the Universe that have some specified property or attribute ("Adjunct" for Carroll). And, since, as we saw, the process of classification is mental, it can be applied to Things whether they exist or not. So, there are no empty classes, classes that have no constituents. For, a class, by Carroll's lights, is simply its constituents (it is not a *set* as modern logicians would see it). Logicians say that this view commits one to the principle of *existential import*, which requires that the subject of any universal proposition must correspond (and refer) to things in the universe of discourse. Consequently, in traditional logic, a universal proposition of the form 'Every A is B' logically entails 'Some A is B' (likewise, 'No A is B' entails 'Some A isn't B'). Modern mathematical logicians allow for empty sets. For example, the set of solar planets that are made entirely of ice cream has no members. It follows that they reject the principle of existential import. And that is an important difference between traditional logic and modern mathematical logic.

But there is more to this story. After writing that "the Universe contains **Things**", Carroll wrote that "Things have **Attributes**." Well, must a thing have attributes? That's one of the questions that philosophers have often found bewildering. Some say yes; some say no. Carroll rejected any answer. In the first chapter of *The Game of Logic*, he wrote this about the question:

> People have asked the question "Can a Thing exist without any Attributes belonging to it?" It is a very puzzling question, and I'm not going to try to answer it: let us turn up our noses, and treat it with contemptuous silence, as if it really wasn't worth noticing.

It is interesting to note here that he immediately turned to the related question of whether an Attribute could exist without any Thing, to which he provided a negative reply – in a way only Lewis Carroll could do: "You never saw 'beautiful' floating about in the air, or littered about on the floor, without any Thing to *be* beautiful, now did you?" Carroll was insightful and correct here. There are no Things without Attributes; no Attributes without Things.

A thing with no attributes is called a *bare particular*. Strictly from the point of view of modern logic, there must be bare particulars. This is how perhaps the most prominent American logician of the Twentieth Century put it:

> The pronoun is the tenable linguistic counterpart of the untenable old metaphysical notion of a bare particular. (Quine 1980, 165)

> The variable is the legitimate latter-day embodiment of the incoherent old idea of a bare particular. (Quine 1981, 25)

Quine might not have liked the old idea of a bare particular (he never said why), but he was happy to accept it in its modern version embodied in the new logic. Traditional logic formulates a proposition such as 'Every unicorn is magical' as a Subject ('every unicorn') referring to all unicorns and a Predicate ('is magical') characterizing the referent(s). Modern logic formulates that proposition as 'Everything, (in the domain of discourse) is such that, if it is a unicorn, then it is magical' (symbolically: $(\forall x)(Ux \supset Mx)$). Notice the occurrences of the pronoun 'it' here, which is, as Quine held, the vernacular version of the logicians' *individual variable* (those tokens of 'x' in the symbolism). So, this can, perhaps more transparently be paraphrased as: 'In the domain of discourse, everything (i.e., every bare particular) has the attribute/property of being magical if it has the property of being a unicorn. 'Some lions are tame' would be paraphrased as: 'In the domain, there is at least one bare particular that has the property of being both a lion and tame'. Remember, those bare particulars, now disguised by pronouns/variables, are *bare* (at least until they are accorded properties, such as being a unicorn or being magical, in due course).

5. "The Actual Facts of Life"

Traditional logicians adhered to the principle (called *subalternation*) that a particular proposition (e.g., 'Some logicians are poets' is immediately entailed by its corresponding universal ('All logicians are poets'). As well, they generally held that a particular proposition implicitly entails that something (in the universe of discourse) is referred to by the subject ('There is a logician'). Consequently, the universal proposition entails the existence of at least one thing to which its subject refers. That is the principle of existential import. Modern logicians reject that principle (and thus they reject subalternation). The key difference revolves around how one is to analyze the logic of universal propositions.

Carroll claimed that a universal proposition "contains" its corresponding particular. He gave (*Symbolic Logic*, Part I, Book II, Chapter III) this example:

> [Thus, the Proposition "*All* bankers are rich men" evidently contains the smaller Proposition "*Some* bankers are rich men."]

And this was a position he held for a long time. Eventually, however, he seemed to have doubts. In Part II of *Symbolic Logic* (Book X, Chapter II), he addressed the issue directly and at some length. There he still held that A propositions contain their corresponding E propositions. But, he also suggested that any theory about this, when "applied to the actual facts of life," must satisfy the test that they not be "singularly inconvenient for ordinary folk." (Bartley 1986, 234). Even in Part I, Carroll hinted that a different approach to the problem might be taken In Part II: "Note that the rules, here laid down, are arbitrary, and only apply to Part I of my *Symbolic Logic*." (Bartley 1986, 76) So, given that he came to believe that a logical thesis should not be inconvenient for ordinary folks when it is applied to "the actual facts of life," what did he come to say about existential import in Part II?

Carroll wrote there that a proposition *asserts* just in case it asserts the existence of its subject. I propositions obviously assert in this way, and so do A propositions, because they *contain* the corresponding I proposition. He then went on to argue that E propositions do not assert, because the assumption that they do leads to a contradiction. Finally, Carroll claims that, one could argue that both A and E propositions assert because any A proposition is logically equivalent to an E proposition (i.e., 'Every X is Y' is equivalent to 'No X is not-Y'). However, that argument would not pass his crucial test (it would be inconvenient and incompatible with the facts of life). The upshot of these considerations is that Carroll believed that A propositions do assert but that the choice then is between taking E propositions to assert or I propositions to assert. He chose to adopt the second choice because of the *inconvenience*, etc. of the first choice. Yet he said that his choice was "the one adopted in this book." It was a policy decision, not necessitated by logic alone. In Note B, at the end of Book X, he appears to suggest that considerations of common linguistic usage are often useful in making such choices. It seems fairly clear that Carroll remained loath to give up the idea of existential import for universal propositions.

6. "Shut Their Eyes Like Frightened Children"

Aristotle was careful to spell out the notion of *opposition*, especially when it comes to terms and propositions (especially in Book X of his *Categories*). In the case of terms, some pairs are opposite in the sense that one or the other must be true of any subject to which they would apply. For example, 'sighted' and 'blind' are such that for any animal (i.e., a thing to which such terms would naturally apply) just one truly holds. By contrast, a rose cannot be said to be either sighted or blind. For Aristotle, a term like 'blind' was a "privative" term. Pairs consisting of a term and its privative admit no intermediaries. Whatever is, by nature, one cannot be the other. They are (logically) *contraries*. However, there are many pairs of terms that cannot both the true of the same thing and the same time, but they have intermediaries that can hold along with one or the other. For example, 'good' and 'bad' are contrary (in the sense that they cannot both simultaneously hold of the same subject, Nevertheless, there are people who are not bad but nonetheless are not necessarily good. They're somewhere in between. Such pairs of terms are contraries – but not logical contraries. In his *On Interpretation* (Chapter II), Aristotle wrote that terms that are privative are "indefinite" terms of the form 'not-x'. So, terms seem to come in pairs, 'x'/ 'not-x'. But so do propositions. Two propositions opposed in this way are *contradictories*. Aristotle's account here became a part of the very foundation of traditional logic.

Carroll, unsurprisingly, accepted this tradition of distinguishing between the negation of a term and the negation of a proposition. Yet many of his contemporary logicians (and ultimately, most of subsequent mathematical logicians) did not. Referring to the former group, he wrote:

> The fact is, "The Logicians" have somehow acquired a perfectly *morbid* dread of negative Attributes, which makes them shut their eyes, like frightened children, when they come across such terrible Propositions as "All not-x are y"" and thus they exclude from their system many very useful forms of Syllogisms. (Bartley 238)

He had written something quite similar at the end of Chapter I of his *The Game of Logic.*

Throughout his works on logic, Carroll was happy to use negative terms, which are, after all, quite common in English. His confidence in recognizing the role of term negation for logic was born of his clear-eyed understanding of the difference between contradictory propositions and negative terms. Early on in *Symbolic Logic* (Book I, Chapter III) he wrote this about the latter:

> Henceforwards let it be understood that, if a Class of Things be divided into two Classes, whose Differentiae have contrary meanings, each Differentia is to be regarded as equivalent to the other with the word "not" prefixed.

In other words, *logically contrary* terms, which he confusingly calls *contradictory* terms (Bartley 285-286), divide a class exclusively and exhaustively – just as Aristotle had taught. In the next chapter, Carroll admits negative terms even when they are in the subject, the very thing that most frightened "The Logicians," showing that, for example, 'None but the brave deserve the fair' is equivalent to 'No not-brave deserve the fair'. He did not go on to recognize that 'no' itself is not, as he thought, a quantifier. It is a portmanteau word, analyzable as 'Not: some/any'.

Modern logicians understand the contradictory of a proposition as the result of applying function to the proposition, a function of *propositional negation.* Carroll, good traditionalist that he was, understood the contradictory of a proposition to be its *denial.* And the denial of a proposition is the result of two things: change of quantity and change of quality. The first change involves making a universal a particular or a particular a universal (i.e., an exchange between A and I or an exchange between E and O). The second change involves exchanging a positive copula for a negative copula. Importantly, a negative copula is not a negative term.

7. "With Bated Breath"

All Carroll tells us about logical a copula in Book II, Chapter I, of *Symbolic Logic* is that it is "The verb 'are' (or 'is'). (This is called the **Copula**.) " Not much more light is shed on the topic when he mentions it again in *Symbolic Logic*, Part II, Book X, Chapter II. Yet it is still worth quoting in full. After all, only Carroll could mention the topic in such a delightful way.

> The writers, and editors, of the Logical text-books which run in the ordinary grooves – to whom I shall hereafter refer by the (I hope inoffensive title "The Logicians" – take on this subject, what seems to me to be a more humble position than is at all necessary. They speak of the Copula of a Proposition "with bated breath," almost as if it were a living, conscious Entity, capable of declaring for itself what it chose to mean, and that we, poor human creatures, had nothing to do but to ascertain *what* was its sovereign will and pleasure, and submit to it.

He took a *slightly* more serious tone in the following chapter, where he directly addressed the question of whether one should, when appropriate, attach a 'not' to the copula or to the predicate. In other words, what, if any, is the logical difference between propositions of the forms 'Some S is-not P' and 'Some S is not-P' (e.g., 'Some head of state isn't competent' and 'Some head of state is incompetent')? His answer:

> When they [The Logicians] are putting the final touches to the grouping of their Propositions, just before the curtain goes up and when the Copula – always a rather fussy "heavy father," asks them "Am *I* to have the 'not,' or will you tack it on to the Predicate?" they are much too ready to answer, like the subtle cab-driver, "Leave it to *you*, Sir!" The result seems to be, that the grasping Copula constantly gets a "not" that had better have been merged in the Predicate, and that Propositions are differentiated which had better have been recognised as precisely similar.

Carroll's choice to treat the two alternative propositional forms as logically equivalent amounts to the traditional rule of immediate inference called *obversion*. In effect, one use of the rule is to eliminate negative copulae in favor of a negative predicate term. As it happens, a system of formal logic that systematically admits both negative copulae and negative terms has important advantages inference (see Sautter 2021, final paragraph). But that's a story for another day.

A brief final note: The work presented here is the direct result of a suggestion, followed by much encouragement, then followed by invaluable input from a first-class historian of Nineteenth Century British logic and Carroll's place there. Without all of that, this essay would not be a Thing existing in the Real, rather than Imaginary, Universe. I thank and dedicate this essay to Amirouch Moktefi.

References

Abeles, F. and Moktefi, A., 2011, "Hugh MacColl and Lewis Carroll: Crosscurrents in Geometry and Logic," *Philosophia Scientiæ: Travaux d'histoire et de philosophie des sciences*, 15: 55-76.

Ackrill, J. L., *Aristotle's* Categories *and* De Interpretatione, Oxford: Clarendon Press.

Bartley, W.W. III, ed., 1986, *Lewis Carroll's Symbolic Logic*, NY:Clarkson N. Potter.

Carroll, Lewis, 1886, *The Game of Logic*, London: MacMillan.

Carroll, Lewis, 1896, *Symbolic Logic. Part I: Elementary*, 1st ed., London: MacMillan.

Carroll, Lewis, 1896, *Symbolic Logic. Part I: Elementary*, 2nd ed., London: MacMillan.

Carroll, Lewis, 1896, *Symbolic Logic. Part I: Elementary*, 3rd ed., London: MacMillan.

Carroll, Lewis, 1897, *Symbolic Logic. Part I: Elementary*, 4th ed., London: MacMillan. Reprinted in Bartley 1986.

Englebretsen, G., 1986, "Czezowski on Wild Quantity, " *Notre Dame Journal of Formal Logic*, 27: 62-65.

Englebretsen, G., 1988, "A Note on Leibniz's Wild Quantity Thesis," *Linguistic Analysis*, 20: 87-89.

Englebretsen, G., 1996, *Something to Reckon With: The Logic of Terms*, Ottawa: University of Ottawa Press.

Leibniz, G.W., 1966, "A Paper on 'Some Logical Difficulties'," *Leibniz: Logical Papers*, G.H.R. Parkinson (ed.), Oxford: Clarendon Press, pp. 115-121.

Prior, A.N., 1957, "Review: Lewis Carroll, *Symbolic Logic. Part I. Elementary*," *Journal of Symbolic Logic*, 22: 309-310.

Quine, W.V.O., 1980, "The Variable and its Place in Reference," *Philosophical Subjects*, Z. van Straaten (ed.), Oxford: Clarendon Press.

Quine, W.V.O., 1981, "Grammar, Truth, and Logic," *Philosophy and Grammar*, S. Kanger and S. Ohman (eds.), Dordrecht: D. Reidel.

Sautter, F.T., 2012, "Termos Negativos e Diagramas," 1er Congreso de la Sociedad Filosófica del Uruguay Simposio Diagramas.

Sommers, F., 1967, "On a Fregean Dogma," *Problems in the* Philosophy *of Mathematics*, I. Lakatos (ed.), Amsterdam: North-Holland Publishing.

Sommers, F., 1969, "Do We Need Identity?" *Journal of Philosophy*, 66: 499-504.

Sommers, F., 1970, "The Calculus of Terms," *Mind*, 79: 1-39.

Sommers, F., 1976a, "Frege or Leibniz?" *Studies on Frege, Vol. III*, M. Schirn (ed.), Stuttgart: Fromann-Holzboog.

Sommers, F., 1976b, "On Predication and Logical Syntax," *Language in Focus*, A. Kasher (ed.), Dordrecht: D. Reidel.

Sommers, F.,1982, *The Logic of Natural Language*, Oxford: Clarendon Press.

Sommers, F. and Englebretsen, G., 2000, *An Invitation to Formal Reasoning*, Aldershot: Ashgate.

Just One More Note

"Begin at the beginning, and go until you come to the end; then stop."

The Kind of Hearts to the White Rabbit

Well, yes, that was good advice from the King. We have come to the end of this series of essays, reviews, and notes. So just consider this brief note as an unforeseen afterthought. I feel the need to say something not about the logical contributions of Lewis Carroll but about how I personally came to be interested in such things in the first place and where that led me.

I grew up in the 1940s and '50s in the middle of the U.S. My family was poor. We owned no books. My mother read the bible and my father read the newspaper sometimes. Yet, I was taught to read before I began elementary school. Soon, my aunt would take me to the local library, where she would check-out books for me to take home and read. One was *Alice in Wonderland*. In short order, I was walking to the library on my own, selecting a book, and reading it there. One day the librarian asked if I would like to borrow books to take home. I became a book worm. I spent as much time as I could in the library. A few years later, I haunted the local bookstore, whose kind owner (a friend of mine for the next 30 years of her life), let me read books (that I couldn't actually buy) whenever I wanted. When I was about 16 years old, she saw me reading a battered copy of *Alice Through the Looking-Glass* and remarked, "So I see you've discovered that Lewis Carroll didn't write only for children." No indeed.

I went off to university, where I started out concentrating on Philosophy, Mathematics, and Literature. I eventually earned degrees in Philosophy, with a special focus on logic and the philosophy of language. And then I became a professor. I lectured, taught, researched, wrote, and published for a long time (even after I retired). However, I never ignored Carroll. I have four children and six grandchildren; the first books I read, and gave, to them were the *Alice* books and *The Hunting of the Snark*. Once or twice my late colleague and dear friend Bill Shearson and I led a seminar course on Metaphysics and Epistemology, and, quite understandably, our main texts were Carroll's tales. But, more than that, I had examined and come to appreciate the work on logic that Carroll did during the final years of his life.

In my own research in logic, I had been primarily interested in two topics: the history of logic and how traditional syllogistic logic had been so rapidly and thoroughly replaced by modern mathematical logic (a logic that I taught for half of a century); and the use of diagrams or graphs in logic. I devoted much of my professional career to the study of traditional logic, eventually working with the incredibly original logician Fred Sommers to develop a system of formal logic that shares many key features with traditional syllogistic logic (naturalness and simplicity) while also doing what the old logic could not do – match the expressive and inferential powers of the new logic. Eventually, through this study, I came to a better appreciation of what I believe was Aristotle's use of diagrams in teaching his syllogisms, and then of the historical development of logic diagrams from Leibniz, Lambert, Euler, Venn, Carroll, and Peirce. I wrote and published a large number of essays and books devoted to spelling out my ideas about traditional logic and about diagrams. And what became quite obvious was that Lewis Carroll, that "obscure Writer on Logic, towards the end of the Nineteenth Century" who was working completely within the final stages of traditional logic's dominance in the schools, had made a number of impressive logical innovations, and developed a powerful system of logic diagrams that went well beyond the standard set by Venn.

Little of this was appreciated during his lifetime. Yet, now, in the 21st century, a growing number of scholars, especially, logicians, have come to appreciate Carroll's work in logic. The two most prominent among them today are Francine Abeles and Amirouche Moktefi. I've tagged along, so this modest collection is the result.

www.ingramcontent.com/pod-product-compliance
Lightning Source LLC
LaVergne TN
LVHW050538100826
845148LV00002B/603

* 9 7 8 1 8 4 8 9 0 3 7 5 3 *